W9-BBU-342

DATE DUE			

HIGHSMITH #45230

Printed
in USA

Cover design by Joseph Sherman, Hamden, Ct.

Nathaniel Grossman

The Sheer Joy
of Celestial Mechanics

Birkhäuser
Boston · Basel · Berlin

Nathaniel Grossman
Department of Mathematics
University of California at Los Angeles
Los Angeles, CA 90024-1555

Library of Congress Cataloging-in-Publication Data

Grossman, Nathaniel, 1937-
    The sheer joy of celestial mechanics / Nathaniel Grossman.
        p.   cm.
    Includes bibliographical references and index.
    ISBN 0-8176-3832-6 (H : acid free paper). -- ISBN 3-7643-3832-6 (H :
acid free paper)
    1. Celestial mechanics.   I. Title.
QB351.G69   1995   95-34467
    521--dc20   CIP

Printed on acid-free paper

Birkhäuser ®

© 1996 Birkhäuser Boston

ISBN 0-8176-3832-6
ISBN 3-7643-3832-6
Reformatted from author's disk by TeXniques, Inc., Boston, MA
Printed and bound by Quinn-Woodbine, Woodbine, NJ
Printed in the U.S.A.

9 8 7 6 5 4 3 2 1

Contents

List of Figures

Preface

Dear Reader,

Here is your book. Take it, run with it, pass it, punt it, enjoy all the many things that you can do with it, but—above all—read it. Like all textbooks, it was written to help you increase your knowledge; unlike all too many textbooks that you have bought, it will be fun to read.

A preface usually tells of the author's reasons for writing the book and the author's goals for the reader, followed by a swarm of other important matters that must be attended to yet fit nowhere else in the book. I am fortunate in being able to include an insightful prepublication review that goes directly to my motivations and goals. (Look for it following this preface.) That leaves only those other important matters.

In preparing the text, I consulted a number of books, chief of which included these:

- S. Chandrasekhar, *Ellipsoidal Figures of Equilibrium,* Yale University Press, 1969.
- J.M.A. Danby, *Fundamentals of Celestial Mechanics,* Macmillan, 1962. Now available in a 2nd edition, 3rd printing, revised, corrected and enlarged, Willmann-Bell, 1992.
- Y. Hagihara, *Theories of Equilibrium Figures of a Rotating Homogeneous Fluid Mass,* NASA, 1970.
- R.A. Lyttleton, *The Stability of Rotating Liquid Masses,* Cam-

bridge University Press, 1953.

- C.B. Officer, *Introduction to Theoretical Geophysics,* Springer-Verlag, 1974.
- A.S. Ramsey, *Newtonian Attraction,* Cambridge University Press, 1949.
- W.M. Smart, *Celestial Mechanics,* Longmans, Green, and Co, 1953.
- E.T. Whittaker, *Analytical Dynamics,* Cambridge University Press, 1927.

Readers familiar with these books will recognize the great debt that I owe to them. Other books have offered both comfort and enjoyment; they are credited in footnotes.

I suggest to instructors that this course be taught without either midterm or final examinations. Those medieval rituals may still be necessary to goad recalcitrant calculus students into studying, but they are no longer needed for upper-division students who, after all, are volunteers. In lieu of the examinations, require students to complete a prespecified number of the problems, exactly which ones being each student's choice. These problems may be handed in singly or in batches at any time during the term, and they must be *essentially correct* in order to receive credit. Problems that are not worked correctly are to be quickly recycled back to the student marked with helpful suggestions for his or her emendation and resubmission. Both the mathematical content and the presentation must be scrutinized. It may be useful to the students to distribute copies of J.J. Price's 'Learning Mathematics Through Writing: Some Guidelines,' *The College Mathematics Journal,* 20 (1989) 393–402, which will help them to write well in the journalese that is the received style nowadays. The effort that they put in to improve their mathematical writing will pay off in all of their writing.

One of the useful byproducts of the xerographic reproduction age is a plentiful supply of scratch paper. Students are often reluctant to make sketches as they study, perhaps because they feel that their sketches are too crude. But even the crudest sketching can be helpful. After all,

Archimedes traced his figures in the sand as he worked. John Aubrey (1626–1697) wrote of the philosopher Thomas Hobbes (1588–1679)[1]

> I have heard Mr. Hobbes say that he was wont to draw lines
> on his thigh and on the sheetes, abed, and also multiply and
> divide.

> *Brief Lives,* edited by O.L. Dick, 1949

Charles G. Lange, my long-time friend and colleague, often urged me to complete this book. His own work was notable for its devotion to real problems arising from the real world, for the beauty of the mathematics he invoked, and for the elegance of its exposition. Chuck died in Summer, 1993, at the age of 51, leaving much undone. Many times since his death I have missed his counsel, and I know that this text would be far less imperfect if I could have asked him about a host of vexing matters. Reader, I hope that you have such a friend as I had.

Nathaniel Grossman
September, 1994

---

[1] Hobbs is a classic example of one who came to mathematics 'late' in life. From Aubrey:

> He was 40 years old before he looked on Geometry; which happened accidentally. Being
> in a Gentleman's Library, Euclid's Elements lay open, and 'twas the 47th El. *libri* 1. He
> read the Proposition. *By G—,* sayd he (he would now and then sweare an emphaticall
> Oath by way of emphasis) *this is impossible !* So he reads the Demonstration of it, which
> referred him back to such a Proposition; which proposition he read. That referred him
> back to another, which he also read. *Et sic deinceps* [and so on] that at last he was
> demonstratively convinced of that trueth. This made him in love with Geometry.

Prepublication Review

The Sheer Joy Of Celestial Mechanics. By Nathaniel Grossman. Ca. 176 pp.

This book presents topics in celestial and vectorial mechanics that have a definite mathematical flavor, with the goal of illustrating how the techniques of lower division calculus classes and of 'advanced calculus' can be put to work. With that goal in mind, the author has not attempted a consistent and complete development of mechanics from first principles, for such a development is available in courses taught in physics, engineering, and other departments. Nor has he included formalistic Lagrangian and Hamiltonian approaches because they are of limited utility to beginners.

It is possible to earn a bachelor's degree in mathematics at many good schools and to learn nothing of those parts of the subject that are the meat and bread of applied mathematicians. Fourier series, Bessel and Legendre functions: if mentioned at all, they get the briefest notice. The Inverse Function Theorem is reduced to a tortured lecture or two in sophomore calculus, never to be encountered again. The beauties of classical potential theory are banished. Even most students receiving the doctorate have not heard of Lagrange's Expansion Theorem. Yet these topics are not obsolete. They are alive and they remain useful. They are not mere armchair exercises. All of these topics are developed and used in this book.

Don't look for axioms to memorize. Too many courses are consecrated to teaching students to play chords on a set of axioms. This book celebrates the heroic age of calculus, the time of Euler, Maclaurin, Clairault, Lagrange, and Laplace, a time before δ and ϵ. Of course, rigor is a glory of modern mathematics, but it is not the crowning glory. Mathematics is concerned with ideas and inspiration, with islands of elegance poking up from a sea of drudgery. Most of all, mathematics was invented to do things, not just to be talked about, and today—still—its greatest triumphs are what it can do.

In this spirit, the author has not put rigor into every place where it belongs. After all, the rigorizing of Fourier series alone accounts for a major part of late nineteenth- and early twentieth-century mathematical endeavor. The aim is to show what calculus can *do*. George F. Simmons[2] states it well: '[The] mere fact that we are not able to seal every crack in the reasoning seems a flimsy excuse for denying students an opportunity to glimpse some of the wonders that can be found in . . . calculus.'

What about the arcane notation of celestial mechanics? 'The use of the special and proper terminology separates the experts from the laymen,' this the dictum of the master celestial-mechanician V.G. Szebehely in his delightful introductory textbook.[3] 'True, a mass, a matins, and a vesper well rung are half said,' as Rabelais[4] put it some four hundred years earlier. While the author is well aware of these and similar adages, and he values their wisdom, he does not always follow their advice. In this book the author is not preaching to the converted—neither does he expect to convert his congregation *en masse* into devoted orbiticians. He expects his readers to know the vernacular of calculus and he addresses them in that vernacular. Having brought them into the cathedral and pointed out to them its celestial beauties, he hopes that some may go on to a deeper service, but knows that all will be moved.

Mathematics in general and celestial mechanics in particular were once

---

[2] G.F. Simmons, *Calculus Gems: Brief Lives and Memorable Mathematics*, McGraw-Hill, Inc., New York, 1992.

[3] V.G. Szebehely, *Adventures in Celestial Mechanics: A First Course in the Theory of Orbits*, University of Texas Press, 1989 (Third Printing, 1993), p. 2.

[4] F. Rabelais, *Gargantua and Pantagruel*, Book I, Chapter XL. Translated by D.M. Frame, University of California Press, 1991.

studied by every educated man. For example,[5]

> [Thomas] Jefferson was familiar not only with current developments in mathematics, but while in Paris [as American ambassador] had become personally acquainted with notable French mathematicians, particularly Joseph Louis Lagrange. Mathematics had always been one of his favorite studies, learned under the tutelage of William Small [at the College of William and Mary], who taught that mathematics was the central hub from which all other sciences branched out. ... [Jefferson] was responsible for giving mathematics a more important place in the curriculum of the University of Virginia than it was given at any other university at that time.

What did Jefferson expect of an education? He declared that[6]

> we do not expect our schools to turn out their alumni enthroned on the pinnacles of their respective sciences; but only so far advanced in each as to be able to pursue them by themselves, and to become Newtons and Laplaces by energies and perseverances to be continued throughout life.

How low our expectations have sunk to where they now rest.

The book under review hopes to shore up a corner of the sagging mathematical education edifice. Here is an outline of its contents—divided into eight chapters—with mention of some of the individual topics.

(1) Rotating Coordinates. Newton's Laws of Motion. Foucault pendulum.

(2) Central Forces. Conservation laws. Integrable cases of power laws. Bonnet's Theorem.

(3) Orbits under the Inverse Square Law. Kepler's equation. Dirac's drifting gravitational constant.

---

[5] S.A. Bedini, *Thomas Jefferson: Statesman of Science*, MacMillan Publishing Company, New York, 1990, p. 322.

[6] Ibid., p. 474.

(4) Expansions Associated to an Elliptic Orbit. Fourier series. Bessel coefficients. Legendre functions. Lagrange's inversion theorem.

(5) Universal Gravitation and Oval Orbits. Bertrand's characterization of Newton's Law of Universal Gravitation. Oval orbits in the large.

(6) Dynamical Properties of Rigid Bodies. Moments of inertia. Eulerian free motion of the Earth. Feynman's wobbling plate. Precession of the equinoxes.

(7) Gravitational Properties of Solid Bodies. MacCullagh's formula. Internal and external potentials of an ellipsoid. Tide-raising by external bodies.

(8) Equilibrium of a Self-gravitating Fluid. Proof of Lichtenstein's theorem that a rotating, homogeneous, self-gravitating fluid body has an equatorial plane of symmetry. Derivation of the Maclaurin and Jacobi ellipsoidal forms of a rotating fluid.

This short list can not do justice to the rich variety of topics.

There are almost one hundred exercises, and they are an important part of the course. Just a few are original. The rest come from well-known books and many have a distinguished pedigree, being traceable back through the generations into prehistory. Many are in the British tradition: They require a stroke of understanding or insight and a felicitous choice of formula but then yield quickly. There are no 'plug and chug' problems, as these are readily available in abundance from many sources. The problems appear in the text at the appropriate place and time. The author suggests working them when they present themselves: They illustrate the points made in the text and, sometimes, extend them. However, only a few of them contain results called upon later in the text.

The reviewer has been close to the author for a long time, and he was pleased when the author announced plans to bring this book into print. An earlier form of the text was developed as notes for a course on celestial mechanics that the author gave at UCLA in Spring, 1977. It drew a modest number of able undergraduates, whose enthusiastic response when they encountered these beautiful applications heartened him. He looked forward to teaching the course again. But circumstances changed, and in recent years the focus of the university administration has changed from presenting advanced undergraduate courses to the talented undergraduates who can appreciate and learn from them into a strategy of core courses for masses of students. The prospect for teachers and students is about as appetizing as a bowl of Shredded Wheat that is to be eaten without milk. The result was and is predictable: The most talented students languish, while the administrators can point to an unabated flow of homogenized 'product,' the result of a 'dumbed-down' curriculum.

Recently the UCLA administrators graciously have allowed professors to teach 'overload' courses for a wretched pittance, although the students will get the customary credit units. It might be argued that this is what professors should have been doing all along, not just for graduate research students but also for undergraduates. Our professorial model is to be the Clerk, of whom Chaucer wrote that 'gladly wolde he lerne, and gladly teche.' The author, for one, jumped at the chance. He has reworked and expanded the notes and will offer the course in Fall, 1994 (and, he tells the reviewer, whenever he can assemble interested students). He finds that presenting beautiful mathematics is rewarding, and that seeing students' pleasure as they encounter this beauty for the first time is greatly rewarding.

Will this book find a wide audience? The reviewer hopes so. He is pessimistic about the future of the infinitesimal calculus course, having seen it decline in both content and expectations in the more than thirty years that he has been teaching calculus. He remains convinced that the calculus is the most powerful mathematical tool ever invented, but in the current state of calculus textbooks, he finds it increasingly difficult to convince students of that fact. He will be elated to have this book in hand, and he anticipates waving it high with the fervor of a sidewalk preacher.

NATHANIEL GROSSMAN
University of California, Los Angeles

1

Rotating Coordinates

1. Some kinematics

We want to look first at some geometric aspects of motion without paying any attention to the causes of motion or to the physical laws that might govern the motion.

In order to cut down the notational clutter and to emphasize the geometrical nature of this material, we will begin our journey along the historic route of infinitesimals. While these objects are indispensable to applied mathematicians and appliers of mathematics, their language has been long banished from the usual calculus books because they were considered to be 'unrigorous.' Thus, recent generations of calculus students have wrestled with δ and ϵ without any idea of how these runic characters affect the *use* of calculus. Indeed, they don't.[1]

The chief operational property of infinitesimals that we use is their infinitesimality. In the first place, any infinitesimal number is supposed to be nonzero, yet smaller in absolute value than any 'ordinary' number. They are 'negligible' in comparison to ordinary numbers. Furthermore,

---

[1]Leibniz's infinitesimals were revived in full mathematical rigor by the logician Abraham Robinson in 1961. They are well described in his book *Non-Standard Analysis,* North-Holland, 1966. You will find an elementary development in H.J. Keisler, *Elementary Calculus,* Prindle, Weber & Schmidt, 1976, but years of precedent make it unlikely that Cauchy's δ-ϵ approach will be displaced.

the square of an infinitesimal and, generally, the product of two infinitesimals is considered to be of 'higher order' and negligible with respect to an 'ordinary' infinitesimal.

Let u be a unit vector, supposed moving in time. Suppose that u moves in time Δt to $u + \Delta u$, where Δu is small, while remaining a unit vector. Then $(u + \Delta u) \cdot (u + \Delta u) = 1$. Neglecting $\Delta u \cdot \Delta u$ as being 'higher order,' there results $u \cdot \Delta u = 0$, so that $\Delta u \perp u$. The motion can be visualized as an infinitesimal rotation[2] around an axis perpendicular to both u and Δu, lying in fact in the direction of $u \times \Delta u$ (Figure I.1). Let k be a unit vector in that direction. Then the direction of Δu is that of $k \times u$ and the magnitude of Δu is $\Delta \vartheta$. This means that $\Delta u = \Delta \vartheta (k \times u)$. Dividing by Δt and letting $\Delta t \to 0$, we get

$$(1) \qquad\qquad \frac{du}{dt} = \omega \times u,$$

where $\omega = d\vartheta/dt\, k$ is the instantaneous *angular velocity* of the motion.

We have written ω as a vector. To justify this, we must prove that it behaves vectorially. Let u undergo successive rotations about the axes k_1 and k_2 by respective angles $\Delta \vartheta_1$ and $\Delta \vartheta_2$. Then the changes in u are first

$$(2) \qquad\qquad u \hookrightarrow u + \Delta u_1 = u + \Delta \vartheta_1\, k_1 \times u,$$

and second to

$$
\begin{aligned}
(3)\; u + \Delta u \;&=\; u + \Delta u_1 + \Delta u_2 \\
&=\; u + \Delta u_1 + \Delta \vartheta_2\,(k_2 \times (u + \Delta u_1)) \\
&=\; u + \Delta \vartheta_1\,(k_1 \times u) + \Delta \vartheta_2\,[k_2 \times (u + \Delta \vartheta_1\,(k_1 \times u))] \\
&=\; u + \Delta \vartheta_1\, k_1 \times u + \Delta \vartheta\, k_2 \times u + \text{second order} \\
&=\; u + (\Delta \vartheta_1 k_1 + \Delta \vartheta_2 k_2) \times u + \text{second order.}
\end{aligned}
$$

To summarize, we get a change

$$(4) \qquad u \hookrightarrow u + (\Delta \vartheta_1 k_1 + \Delta \vartheta_2 k_2) \times u + \text{second order.}$$

---

[2] We take a closer look at rotations in §VI. 8.

But this can be written as

(5) $$\omega = \omega_1 + \omega_2,$$

where

(6) $$\omega_1 = \frac{d\vartheta_1}{dt}\,\boldsymbol{k}_1, \qquad \omega_2 = \frac{d\vartheta_2}{dt}\,\boldsymbol{k}_2,$$

and ω is the resultant angular velocity. It is also true that $\omega_1 + \omega_2 = \omega_2 + \omega_1$, so that *infinitesimal* rotations commute. This is not true for rotations in general.

EXERCISE I.1. Try exercise only if you have taken a course in linear algebra and are familiar with matrices and linear transformations.

Formulate the above in terms of linear transformations and matrices (rotations are orthogonal matrices, etc.). Then obtain the law of addition for angular velocities by using derivatives instead of appealing to 'higher-order' quantities.

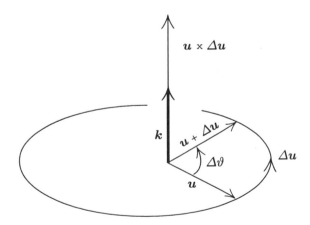

Figure I.1. Infinitesimal rotation

Now we will consider rotating axes in general. There will be a fixed Euclidean framework of orthogonal vectors (the 'inertial' frame). Take orthonormal vectors $\boldsymbol{i}, \boldsymbol{j}, \boldsymbol{k}$ that move as functions of time. Let $\boldsymbol{A}$ be a time-dependent vector and let it have components A_x, A_y, A_z with respect to the moving frame:

(7) $$\boldsymbol{A} = A_x\boldsymbol{i} + A_y\boldsymbol{j} + A_z\boldsymbol{k}.$$

Then

(8) $\quad \dfrac{d\mathbf{A}}{dt} = \dfrac{dA_x}{dt}\,\mathbf{i} + \dfrac{dA_y}{dt}\,\mathbf{j} + \dfrac{dA_z}{dt}\,\mathbf{k} + A_x\,\dfrac{d\mathbf{i}}{dt} + A_y\,\dfrac{d\mathbf{j}}{dt} + A_z\,\dfrac{d\mathbf{k}}{dt}.$

It is possible to show (see Exercise I.2) that there is a *single* angular velocity vector $\boldsymbol{\omega}$ such that

(9) $\qquad\qquad \dfrac{d\mathbf{i}}{dt} = \boldsymbol{\omega}\times\mathbf{i}, \quad \dfrac{d\mathbf{j}}{dt} = \boldsymbol{\omega}\times\mathbf{j}, \quad \dfrac{d\mathbf{k}}{dt} = \boldsymbol{\omega}\times\mathbf{k}.$

We can write therefore that

(10) $\qquad\qquad\qquad \dfrac{d\mathbf{A}}{dt} = \dfrac{\partial\mathbf{A}}{\partial t} + \boldsymbol{\omega}\times\mathbf{A},$

where $\partial/\partial t$ stands for differentiation of the components with respect to time as if $\mathbf{i}, \mathbf{j}, \mathbf{k}$ were fixed.

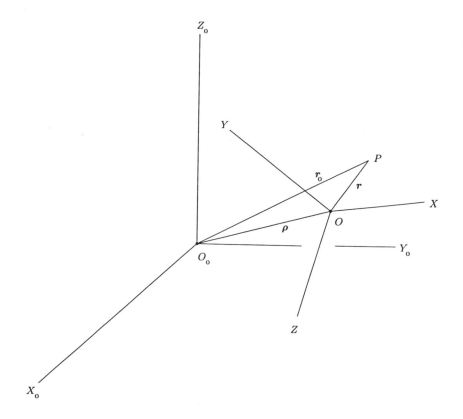

Figure I.2. Fixed and moving frames

Take a fixed coordinate system (subscript$_0$) and let a second coordinate system move in space—the origin O is translating and the axes XYZ are rotating (Figure I.2). Let ρ be the position of O with respect to O_0 and r_0 that of P, while r is the position of P with respect to O. Then $r_0 = \rho + r$ and the velocities satisfy

$$(11) \qquad v_0 = \frac{d\rho}{dt} + v + \omega \times r \qquad \left(v = \frac{\partial r}{\partial t} \right).$$

Repeat the procedure to obtain the equations for acceleration. This gives

$$
\begin{aligned}
(12) \quad a_0 &= \frac{d^2\rho}{dt^2} + \frac{dv}{dt} + \frac{d(\omega \times r)}{dt} \\
&= \frac{d^2\rho}{dt^2} + \frac{\partial v}{\partial t} + \omega \times v + \frac{d\omega}{dt} \times r + \omega \times \frac{dr}{dt} \\
&= \frac{d^2\rho}{dt^2} + a + \omega \times v + \frac{d\omega}{dt} \times r + \omega \times \left(\frac{\partial r}{\partial t} + \omega \times r \right) \\
&= \frac{d^2\rho}{dt^2} + a + 2\omega \times v + \frac{d\omega}{dt} \times r + \omega \times (\omega \times r).
\end{aligned}
$$

The five terms have the following interpretations:

$\dfrac{d^2\rho}{dt^2}$:   The acceleration of the moving origin with respect to the fixed origin and frame.

a: The acceleration of the moving point with respect to the moving frame.

$2\omega \times v$:   The so-called *Coriolis acceleration*.

$\dfrac{d\omega}{dt} \times r$:   The linear acceleration due to the angular acceleration of the moving axes.

$\omega \times (\omega \times r)$:   The *centripetal acceleration* due to the rotation of the moving axes.

Some terms disappear from the formula under certain conditions. For example, suppose P to be rigidly attached to the moving frame. Then v

and a are identically 0, and

$$(13) \qquad a_0 = \frac{d^2\rho}{dt^2} + \frac{d\omega}{dt} \times r + \omega \times (\omega \times r).$$

Note also that the Coriolis acceleration vanishes if v is parallel to ω. (Imagine this on the rotating Earth moving in its orbit through space.)

When the axes are fixed in the Earth and are issuing from its center, then $\frac{d\omega}{dt} = 0$,[3] and

$$(14) \qquad a_0 = \frac{d^2\rho}{dt^2} + a + 2\,\omega \times v + \omega \times (\omega \times r).$$

In general, for motions on the Earth, $\frac{d^2\rho}{dt^2}$ can be neglected in comparison to the other terms.

EXERCISE I.2. Let $R(t)$ be a 3×3 matrix function such that $R(t)$ is an orthogonal matrix for all t.

(a) Show that $R'(t)R(t)^T$ is a skew symmetric matrix.

(b) Show that $R(t)$ has either one or three real eigenvalues and that, in the first case, $R(t)$ causes instantaneous rotation about the real eigenvector.

(c) Relate the angular velocity vector ω implied in part (b) to the elements of the skew symmetric matrix in part (a).

Now we consider motion restricted to a plane. Let i point in the *outward* radial direction (from O to P) and let j point orthogonally in the counterclockwise sense of rotation from i (Figure I.3).

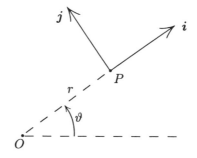

Figure I.3. Plane frame

---

[3]In reality, both the magnitude and direction of the Earth's angular velocity vector vary by small but measurable amounts with time.

Then $k = i \times j$ and $r = ri$. In calculating, we translate the frame from P to O, so $O = O_0$ and, in our earlier notation, $\rho = 0$ and $r = r_0$.

Differentiating gives

$$\text{(15)} \qquad \frac{dr}{dt} = \frac{dr}{dt}\, i + r\, \frac{di}{dt}.$$

From the figure, $i = (\cos \vartheta, \sin \vartheta)$ and $j = (-\sin \vartheta, \cos \vartheta)$, so that

$$\text{(16)} \qquad \frac{di}{dt} = \frac{d\vartheta}{dt}\, j.$$

Hence

$$\text{(17)} \qquad \frac{dr}{dt} = \frac{dr}{dt}\, i + r\frac{d\theta}{dt}\, j.$$

Repeating the differentiation gives

$$\text{(18)} \qquad \frac{d^2 r}{dt^2} = \left(\frac{\partial}{\partial t} + \frac{d\vartheta}{dt}\, k \times\right)\left(\frac{dr}{dt}\, i + r\frac{d\vartheta}{dt}\, j\right)$$

$$= \left[\frac{d^2 r}{dt^2} - r\left(\frac{d\vartheta}{dt}\right)^2\right] i + \left[r\frac{d^2\vartheta}{dt^2} + 2\frac{dr}{dt}\frac{d\vartheta}{dt}\right] j.$$

Thus, the acceleration can be resolved into these components:

$$\text{radial:} \quad \frac{d^2 r}{dt^2} - r\left(\frac{d\vartheta}{dt}\right)^2$$

$$\text{transverse:} \quad r\frac{d^2\vartheta}{dt^2} + 2\frac{dr}{dt}\frac{d\vartheta}{dt}$$

Note that the transverse component of acceleration can be written as

$$\text{(19)} \qquad \frac{1}{r}\frac{d}{dt}\left(r^2\frac{d\vartheta}{dt}\right).$$

The term in parentheses is twice the *areal velocity* (Figure I.4):

$$\text{(20)} \qquad \frac{dA}{dt} = \tfrac{1}{2} r^2\frac{d\vartheta}{dt}.$$

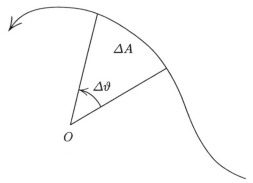

Figure I.4. Areal velocity

For example, suppose that a particle moves in a circle centered at the origin with constant speed. Then dA/dt is constant and $dr/dt = 0$. Thus,

$$(21) \qquad \frac{d^2 \boldsymbol{r}}{dt^2} = -r \left(\frac{d\vartheta}{dt} \right)^2 \boldsymbol{i}.$$

(This corresponds to the term $\boldsymbol{\omega} \times (\boldsymbol{\omega} \times \boldsymbol{r})$ in the general formulation of (12), with $\boldsymbol{\omega} = (d\vartheta/dt)\boldsymbol{k}$.) An observer sitting on the particle and using the moving frame $(\boldsymbol{i}, \boldsymbol{j})$ will find his mechanics wrong unless he introduces a fictitious 'centrifugal' acceleration $-r \left(\dfrac{d\vartheta}{dt} \right)^2 \boldsymbol{i}$.

EXERCISE I.3. A particle moves in an ellipse with semi-axes a and b, with constant areal velocity dA/dt around the center of the ellipse. Find the components of its velocity in Cartesian and polar coordinates. Also find the period of the orbit.

2. Dynamics

In §1, we used 'kinematics' to mean the study of geometric properties of the motion of a particle (and, in general, of a system of particles) without taking into account any causes of the motion or any interactions among the several particles in a system. Thus, kinematics is a branch of mathematics. When causes and mutual interactions are taken into account, the study of motion is called 'kinetics.' When special attention is paid to force as the cause of motion, the science is called 'dynamics.'

We will reprise the basic definitions, which we will use at about the level of freshman physics. We assume knowledge at that level. There is, after all, a naive level of understanding that suffices for our needs. Most of the basic concepts are very complicated from a philosophical viewpoint. We will accept mechanics as an *experimentally* based science

Thus, there are *forces* that behave vectorially. There are *mass, length, time.* A useful fiction is the notion of a *point mass* or *particle.* (Later we will investigate whether ordinary bodies can be well represented by point masses.) There is also a set of absolute reference frames, the so-called *inertial frames,* whose specification is one of the tasks of astronomy. All vectorial quantities ultimately will be referred to an inertial frame. The product of the mass m and the velocity v of a moving body is its *linear momentum, mv.* Suppose that the mass is at position r with respect to the origin O of an inertial frame. Then the cross-product $r \times mv$ is called the *angular momentum* with respect to O (or, sometimes, the *moment of momentum*). If the mass is constant in time, then (cf. (17))

$$(22) \qquad r \times mv = (ri) \times m \left(\frac{dr}{dt} i + r \frac{d\vartheta}{dt} j \right) = mr^2 \frac{d\vartheta}{dt} k$$

in the moving polar frame. Hence, the angular momentum is a vector quantity, perpendicular to the plane of motion (that spanned by r and v) and of length equal to $2m$ times the areal velocity.

The *kinetic energy* of the particle is $\frac{1}{2}m|v|^2$, a scalar quantity.

If the constant force F acts upon a particle, causing a displacement r, then the *work* done by the force is $F \cdot r$, a scalar quantity. If the particle moves from point A to point B along a certain curve, and if it is acted upon by a force F that is a function only of the position of the particle on the curve, then the work is calculated as the sum of contributions from the many infinitesimal portions of the curve. The total work is expressed as

$$(23) \qquad W_{AB} = \int_A^B F \cdot dr,$$

where it is understood that the line integral is taken along the curve. The work W_{AB} may depend upon the choice of the curve from A to B.

In doing calculations, all quantities must be expressed in some consistent system of units. Then equations can be checked in a rough way by noting that they must be dimensionally homogeneous.

Now let $F = F(r)$ be a *force field:* For every vector r terminating in a certain region of space, there is a vector $F(r)$. Generally, F is continuous and even continuously differentiable, but there can be exceptions.

Here is one way to get a force field. Let $V(r)$ be a scalar function defined and differentiable in some region. Then there is a vector field given by the negative gradient of V:

$$(24) \qquad\qquad F(r) = -\nabla V(r).$$

(The negative sign is put in for later convenience.) Such a function V is called a *potential energy* function. If a V exists for a given F, it certainly is not unique, since then $F = -\nabla(V + c)$ for any constant function c. Not every F has a potential function V. For a potential energy function to exist, it is necessary that F be *irrotational:* $\nabla \times F = 0$. If F does have a vanishing curl, then V exists *locally.*

If a potential energy function V exists for F, the field F is called *conservative.* In this case, the work in going from A to B along a given path[4] is

$$(25) \qquad\qquad \begin{aligned} W_{AB} &= \int_A^B (-\nabla V) \cdot dr \\ &= -\int_A^B \nabla V \cdot t \, ds \\ &= -\int_A^B \frac{dV}{ds} \, ds \\ &= -[V(B) - V(A)]. \end{aligned}$$

Thus, the work depends only upon the endpoints of the path. Conversely, it follows from Stokes's Theorem that whenever the work depends only upon the path, the force field is conservative.

In problems of celestial mechanics, often the arbitrary constant is chosen so that the potential energy has limit zero as $|r| \to \infty$.

The surfaces $V = $ constant are called *level surfaces* or *equipotentials.* No work is done by displacement along a curve lying entirely

---

[4] $t = dr/ds$ is the tangent vector, ds is the arc length, and $|t| = 1$. dV/ds is the directional derivative of V along the curve.

on a single equipotential surface. This is because ∇V is perpendicular to the surface: $\nabla V \cdot t = 0$ for any curve on the surface, so that $V(B) - V(A) = \int_A^B \nabla V \cdot t \, ds = 0$.

3. Newton's Laws of Motion

The passage from kinetics to dynamics is effected through *Newton's Laws of Motion:*

   (1) Every body perseveres in its state of rest or of moving uniformly in a straight line, except insofar as it is made to change that state by external forces.
   (2) Change of motion is proportional to the impressed force and takes place in the direction in which the force is impressed.
   (3) Reaction is always equal and opposite to action; that is to say, the actions of two bodies upon each other are always equal and in opposite directions.

Taken as mathematical axioms, these laws lead to a wonderfully developed and beautiful subject called *classical mechanics* or *rational mechanics.* Whether that mathematical subject has anything to do with the real world depends upon whether the validity of Newton's Laws can be checked by experiment and, in fact, by a diversity of experiments. They have been checked and are considered quite satisfactory for the major part of the demands made upon them. Certain phenomena can not be explained satisfactorily by Newton's Laws at the present time—one example, bearing directly upon Celestial Mechanics, is the observed advance in the planet Mercury's perihelion.

There are certain aspects of Newton's Laws that lead to philosophical problems. We have already mentioned the requirement of an inertial frame that is needed to describe the state of absolute rest or uniform motion envisaged in the First Law. Following the successes of the wave theory of light and of the mathematical theory of elasticity in the early part of the nineteenth century, it was widely believed that space might be filled with some medium (called the *aether*) through which heavenly bodies passed without leaving a wake and in which light waves could propagate just as waves propagate in elastic bodies. However, the famous Michelson-Morley experiment, performed in stages through the 1880s, showed that no such medium exists. There is furthermore a question of the measurement of time involved in describing uniform motion.

Mathematically, there is no problem. Space is three-dimensional Euclidean space and time is the real number line. A motion is uniform precisely when it is described by the function $r(t) = r_0 + tv$ for constant vectors r_0 and v, and the particle is at rest when $v = 0$.

In the Second Law, Newton called 'motion' what is now called *momentum*. By 'impressed force' he meant what is now called *impulse*. In the simplest case, a constant force F acting for a time Δt imparts an impulse $F\Delta t$, a vector quantity. Then the Second Law can be rephrased:

> The change of momentum of a particle is numerically equal to the impulse that produces it, and is in the same direction.

4. The Laws of Motion and conservation laws

The momentum of a particle of mass m and velocity v is mv. Suppose the particle to be moving as a result of a field of force F. If at a certain time the particle is at $r = r(t)$, then we can approximate the impulse over a subsequent small time interval Δt by $F(r)\Delta t$. By the Second Law, the change in momentum is $\Delta(mv) \cong F(r)\Delta t$. The approximation will get better as $\Delta t \to 0$ provided that F is a sufficiently smooth vector field. After dividing by Δt and letting $\Delta t \to 0$, we obtain the Second Law in the form

$$(26) \qquad \frac{d(mv)}{dt} = F.$$

In general, the force field may be a function of time, position, velocity, and conceivably of higher time derivatives, as well as of space derivatives. In real life, the differential equation takes the form

$$(27) \qquad \frac{d(mv)}{dt} = F(t, r(t), v(t)),$$

which is a second-order differential equation for the unknown function $r(t)$. As they arise in practice, such equations usually, but not always, have a unique solution $r(t)$ which takes on specified values $r_0 = r(t_0)$ and $v_0 = v(t_0)$ at a given time t_0.

Returning to the equation

$$(28) \qquad \frac{d(mv)}{dt} = F,$$

we integrate to get

$$(29) \qquad mv = \text{constant} + \int F \, dt,$$

which is the equation of motion at the momentum level. The integral must be interpreted with care. Along the solution curve $r = r(t)$, we evaluate $v(t) = dr(t)/dt$. Then the integral is $\int F(t, r(t), v(t)) \, dt$.

Now suppose that m is constant. We observe first that

$$(30) \qquad \frac{d(r \times v)}{dt} = \frac{dr}{dt} \times v + r \times \frac{dv}{dt} = r \times \frac{dv}{dt}$$

because $v \times v = 0$. It follows that

$$(31) \qquad \frac{d(r \times mv)}{dt} = r \times F.$$

The quantity $r \times F$ is called the *moment of force* or the *torque* of F with respect to the origin O. It follows therefore from the Second Law that

> the rate of change of angular momentum with respect to
> time is the torque.

EXERCISE I.4. Show that the following are true:

(a) The angular momentum is constant if and only if the motion takes place in a plane and the areal velocity is constant.

(b) The angular momentum is constant if and only if the line of action of the forces always passes through the origin.

Now suppose that the force field F is conservative, with potential energy function V. Then the work done by moving from A to B along a certain curve is

$$(32) \qquad \begin{aligned} W_{AB} &= \int_A^B F \cdot dr = \int_A^B \frac{d(mv)}{dt} \cdot dr \\ &= \int_A^B \frac{d(mv)}{dt} \cdot v \, dt \\ &= \tfrac{1}{2} m |v_B|^2 - \tfrac{1}{2} m |v_A|^2, \end{aligned}$$

if m is constant. From (25), $W_{AB} = -[V(B)-V(A)]$, whence, in moving from A to B,

(33) $$\Delta(\tfrac{1}{2}m|v|^2) + \Delta V = 0.$$

This is the famous theorem of the *conservation of energy.* It can be put into the form

(34) $$\tfrac{1}{2}m|v_A|^2 + V(A) = \tfrac{1}{2}m|v_B|^2 + V(B).$$

The function $\tfrac{1}{2}m|v|^2 + V$ is called the *total energy.* The theorem of conservation of energy states that the total energy function remains constant during the motion (provided that the mass does not change and the force is conservative).

Now suppose that the axes are rotating with angular velocity $\boldsymbol{\omega} = \omega\boldsymbol{k}$, where $\boldsymbol{k}$ always points along the fixed z-axis. We assume ω to be constant. Apply the apparatus of §1, especially the equation

(35) $$\boldsymbol{a}_0 = \frac{d^2\boldsymbol{\rho}}{dt^2} + \boldsymbol{a} + 2\boldsymbol{\omega}\times\boldsymbol{v} + \boldsymbol{\omega}\times(\boldsymbol{\omega}\times\boldsymbol{r})$$

with $\boldsymbol{\rho} = 0$ and $\boldsymbol{\omega} = \omega\boldsymbol{k}$. Then, when $\boldsymbol{r}$ refers to the *moving* axes, the equation of motion is

(36) $$m\left[\frac{d^2\boldsymbol{r}}{dt^2} + 2\omega\boldsymbol{k}\times\frac{d\boldsymbol{r}}{dt} + \omega^2\boldsymbol{k}\times(\boldsymbol{k}\times\boldsymbol{r})\right] = \boldsymbol{F}.$$

Resolve $\boldsymbol{r}$ as $\boldsymbol{r} = \boldsymbol{R} + z\boldsymbol{k}$, where $\boldsymbol{R} \perp \boldsymbol{k}$. Then

(37) $$\boldsymbol{k}\times(\boldsymbol{k}\times\boldsymbol{r}) = \boldsymbol{k}\times(\boldsymbol{k}\times\boldsymbol{R}) = (\boldsymbol{k}\cdot\boldsymbol{R})\boldsymbol{k} - (\boldsymbol{k}\cdot\boldsymbol{k})\boldsymbol{R} = -\boldsymbol{R}.$$

The equation of motion (36) becomes

(38) $$m\left[\frac{d^2\boldsymbol{r}}{dt^2} + 2\omega\boldsymbol{k}\times\frac{d\boldsymbol{r}}{dt} - \omega^2\boldsymbol{R}\right] = \boldsymbol{F}.$$

Form the dot-product of (38) with $d\boldsymbol{r}/dt$. The term with the cross-product drops out. Because the dot-product commutes, we get

(39) $$m\left[\frac{d^2\boldsymbol{r}}{dt^2}\cdot\frac{d\boldsymbol{r}}{dt} - \omega^2\boldsymbol{R}\cdot\frac{d\boldsymbol{R}}{dt}\right] = \boldsymbol{F}\cdot\frac{d\boldsymbol{r}}{dt}.$$

If F has a potential energy function V described with respect to the *moving* axes, we get a *modified energy integral*

$$(40) \qquad \tfrac{1}{2}m|v|^2 + V - \tfrac{1}{2}m\omega^2|R|^2 = \text{constant.}$$

This is the same as the energy integral of motion in the force field with potential energy function $V - \tfrac{1}{2}m\omega^2|R|^2$, called the *modified potential energy*. The term $-\tfrac{1}{2}m\omega^2|R|^2$ is called the *rotational potential energy*.

Note that some authors use potential energy per unit mass. When we do this later, we will refer to the *potential*. Thus, 'potential' and 'potential energy' have different physical units.

It is also important to be aware that some authors relate potential energy to force through the equation $F = \nabla V$.

EXERCISE I.5. Six particles, each of mass m, are placed symmetrically on the circumference of a circle of radius r and are allowed to move from rest under their mutual attractions, which are proportional to the inverse squares of their mutual distances. Find the velocity of each particle when all particles are on the circumference of a circle of radius $\tfrac{1}{2}r$.

5. Simple harmonic motion

A particle moves in a straight line under the influence of a force that varies directly with the distance to a fixed point on the line and attracts to that point. Take the point to be the origin and let x be the (signed) distance to the point from the origin. The force *per unit mass* can be written $-k^2x$, and the equation of motion is[5]

$$(41) \qquad \ddot{x} = -k^2 x.$$

The field of force is conservative with potential $\tfrac{1}{2}k^2x^2$, so that the total energy per unit mass is constant:

$$(42) \qquad \tfrac{1}{2}\dot{x}^2 + \tfrac{1}{2}k^2x^2 = \text{constant.}$$

---

[5]We use Newton's convenient notation. Time derivatives are denoted by superdots. Thus, $\dot{x} = dx/dt$, $\ddot{x} = d^2x/dt^2$, and so forth. The notation becomes typographically inconvenient with three or more dots. For handwritten readability, it is convenient to expand the dots into little circles: $^\circ$.

The constant must be nonnegative (and positive if the particle is to leave the origin), so we write it as $\frac{1}{2}k^2a^2$. The differential equation is well known, and we can write its general solution in the form

$$(43) \qquad\qquad x = a\cos(kt + b),$$

where a and b are constants.

Note that a great deal of information is available from the energy equation without performing a second integration. The energy equation (per unit mass) is

$$(44) \qquad\qquad \tfrac{1}{2}\dot{x}^2 + \tfrac{1}{2}k^2x^2 = \tfrac{1}{2}k^2a^2.$$

The critical value (maximum or minimum) of x occurs when $\dot{x} = 0$; that is, $x_{\text{critical}} = \pm a$. The maximum or minimum of $\dot{x}$ occurs when $\ddot{x} = 0$; from the equation of motion (41), this happens when $x = 0$. The energy equation (44) yields $\dot{x}_{\text{critical}} = \pm ka$. Since only squares occur in the energy equation, there is a lot of symmetry in the motion.

EXERCISE I.6. A small-diameter smooth cylindrical hole is bored through a solid uniform sphere. Suppose that the force on a particle within the sphere is directed toward the center with magnitude proportional to the distance from the center. Show that a particle set free at the surface will oscillate within the hole with a period independent of the direction of the hole.

6. Linear motion in an inverse square field

You undoubtedly will have solved the equations of motion for a particle in a uniform field of force $F = -u$, where u is everywhere self-parallel (pointing 'up'). These equations describe the motion of a projectile travelling near the earth's surface. Away from the surface, the uniform field is no longer a satisfactory approximation to the gravitational force on a particle. According to Newton's Law of Universal Gravitation, to be discussed in detail later, the proper force per unit mass is

$$(45) \qquad\qquad F = -\frac{k^2}{r^2}\, i,$$

where k is a constant, i points radially from the center of the Earth, and r is the distance from the center. Pick a fixed direction and suppose the particle to move on the line in that direction through the center of the Earth. The equation of motion for a particle of unit mass is

$$(46) \qquad \ddot{r} = -\frac{k^2}{r^2}.$$

The potential is $-k^2/r$, so the energy equation can be written

$$(47) \qquad \tfrac{1}{2}\dot{r}^2 - \frac{k^2}{r} = \text{constant}.$$

Let the radial distance be r_0 and the velocity be v_0 when $t = 0$. Then the constant can be evaluated, and the conservation of energy (per unit mass) takes the form

$$(48) \qquad \tfrac{1}{2}\dot{r}^2 - \frac{k^2}{r} = \tfrac{1}{2}v_0^2 - \frac{k^2}{r_0}.$$

Suppose that both $r_0 > 0$ and $v_0 > 0$. We will harvest information from the energy equation (48), which we rewrite as

$$(49) \qquad \dot{r}^2 = \frac{2k^2}{r} + v_0^2 - \frac{2k^2}{r_0}.$$

There are three cases.

Case 1. $v_0^2 < 2k^2/r_0$. The right-hand side of (49) is > 0 when $r = r_0$ and is < 0 when $r = \infty$. Hence, there is an r_1 such that $\dot{r} = 0$ when $r = r_1$. From the differential equation (46), r has a maximum when $r = r_1$, so that r increases from r_0 to r_1 and then decreases.

Case 2. $v_0^2 > 2k^2/r_0$. Then $\dot{r}^2 > v_0^2 - 2k^2/r_0 = \alpha^2 > 0$. Hence, $\dot{r} > \alpha$ for all t, and the particle moves steadily toward ∞ at a speed tending to α as $t \to \infty$.

Case 3. $v_0^2 = 2k^2/r_0$. Then $\dot{r} \to 0$ as $r \to \infty$. This value of v_0 is the *escape velocity*.

7. Pendulum in a uniform gravitational field

We consider an ideal simple pendulum consisting of a bob of mass m, joined by a rigid, weightless rod of length l to a pivot at O. (See Figure I.5.) A uniform force field $\mathbf{F} = -g\mathbf{i}$ per unit mass acts upon the bob. We suppose that the pendulum swings in a fixed plane containing the direction i. By Newton's First Law, it is necessary to involve the tension per unit mass, $-(T/l)r$, acting in the rod.

Let ϑ be the angle from the i-direction (down) to the rod. We use the equations of motion in the angular momentum-torque form (31). Note that the velocity v is always tangent to the circle on which the bob travels. Hence, both sides of (31) represent a vector perpendicular to the plane containing the motion. Of course, $r \perp v$. In fact,

$$(50) \qquad r = l \cos \vartheta\, i + l \sin \vartheta\, j,$$

so that

$$(51) \qquad v = -l \sin \vartheta\, \dot\vartheta\, i + l \cos \vartheta\, \dot\vartheta\, j$$

and

$$(52) \qquad r \times v = l^2 \dot\vartheta\, k.$$

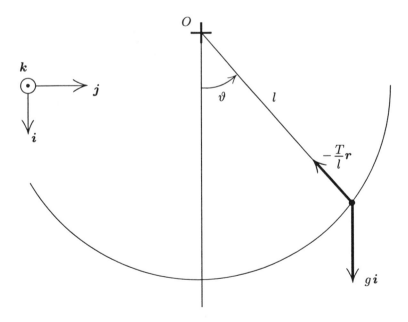

Figure I.5. Simple pendulum

The gravitational force per unit mass is $g\boldsymbol{i}$, so that the associated torque is

$$(53) \qquad \boldsymbol{r} \times \boldsymbol{F} = (l \cos \vartheta \, \boldsymbol{i} + l \sin \vartheta \, \boldsymbol{j}) \times (g\boldsymbol{i} - (T/l)\boldsymbol{r})$$
$$= -gl \sin \vartheta \, \boldsymbol{k}.$$

Considering a unit mass, we get the equation of motion (by equating coefficients of $\boldsymbol{k}$) in the form

$$(54) \qquad \frac{d}{dt}(l^2\dot\vartheta) = -gl \sin \vartheta,$$

which can be rewritten as

$$(55) \qquad \ddot\vartheta + \frac{g}{l} \sin \vartheta = 0.$$

The energy equation follows upon multiplication of (55) by $2\dot\vartheta$ and integration. Supposing that $\vartheta = \vartheta_0$ and $\dot\vartheta = \dot\vartheta_0$ at $t = 0$, we get after simple algebra the relation

$$(56) \qquad \dot\vartheta^2 - \frac{2g}{l} \cos \vartheta = \dot\vartheta_0^2 - \frac{2g}{l} \cos \vartheta_0.$$

After sliding the time scale or rotating the frame vectors a half-turn around $\boldsymbol{i}$, if necessary, we can assume that $\vartheta_0 > 0$. There are three cases to consider.

Case 1. $2g/l = \dot\vartheta_0^2 - (2g/l) \cos \vartheta_0$. The pendulum will, after infinite time, come to rest in a vertically upward position.

Case 2. $2g/l < \dot\vartheta_0^2 - (2g/l) \cos \vartheta_0$. The pendulum will circle endlessly in the same direction because $\dot\vartheta \neq 0$. Because $\dot\vartheta$ depends only upon ϑ, the motion is periodic. However, calculation of the period leads past the elementary functions—as in the next case.

Case 3. $2g/l > \dot\vartheta_0^2 - (2g/l) \cos \vartheta_0$. Then $\dot\vartheta$ will vanish, say for $\vartheta = \alpha > 0$. By the evenness of the energy equation, together with the opposed signs of ϑ and $\ddot\vartheta$, the pendulum will oscillate back and forth between $\vartheta = \alpha$ and $\vartheta = -\alpha$. The energy equation can be rewritten as

$$(57) \qquad \dot\vartheta^2 - \frac{2g}{l} \cos \vartheta = -\frac{2g}{l} \cos \alpha.$$

It is worthwhile to continue with the solution of (57). To be definite, suppose that the bob is drawn aside to $\vartheta = \alpha$ and, at $t = 0$, is released.

Because ϑ will initially decrease as t increases, the negative square root of $\dot{\vartheta}^2$ must be taken, leading upon integration to

$$(58) \qquad t = -\int_\alpha^\vartheta \frac{d\vartheta}{\sqrt{(2g/l)(\cos\vartheta - \cos\alpha)}}.$$

Use the identity $\cos x = 1 - 2\sin^2 \frac{1}{2}x$ on each term under the square root, then introduce a new variable φ by the relation $\sin\frac{1}{2}\vartheta = \sin\frac{1}{2}\alpha \sin\varphi$. The integral transforms as follows:

$$(59) \qquad \begin{aligned} t &= -\frac{1}{2}\sqrt{\frac{l}{g}} \int_\alpha^\vartheta \frac{d\vartheta}{\sqrt{\sin^2\frac{1}{2}\alpha - \sin^2\frac{1}{2}\vartheta}} \\ &= -\sqrt{\frac{l}{g}} \int_{\pi/2}^\varphi \frac{d\varphi}{\sqrt{1 - \sin^2\frac{1}{2}\alpha \sin^2\varphi}}. \end{aligned}$$

We are led therefore to consider the function

$$(60) \qquad F(\varphi, k) = \int_0^\varphi \frac{d\psi}{\sqrt{1 - k^2\sin^2\psi}}, \qquad 0 \le k \le 1,$$

the so-called *incomplete elliptic integral of the first kind.*[6] The parameter k is called the *modulus.* The motion of the pendulum can be written as

$$(61) \qquad t = F(\tfrac{1}{2}\pi, \sin\tfrac{1}{2}\alpha) - F(\varphi, \sin\tfrac{1}{2}\alpha),$$

and in terms of ϑ as

$$(62) \qquad t = K(\sin\tfrac{1}{2}\alpha) - F\left(\sin^{-1}\left[\frac{\sin\frac{1}{2}\vartheta}{\sin\frac{1}{2}\alpha}\right], \sin\tfrac{1}{2}\alpha\right).$$

We have followed standard usage by setting $K(k) = F(\frac{1}{2}\pi, k)$. This is the *complete elliptic integral of the first kind.* To obtain ϑ as an explicit function of t, we must 'invert' the elliptic integral. This is done in terms of elliptic functions, which bear the same relation to the elliptic integrals as the circular function sine does to the integral $\int_0^x (1-u^2)^{-1/2}\, du$. The theory of elliptic functions is complicated, but those problems whose solutions can be expressed in terms of elliptic functions are considered 'elementary' by all but the most recent writers on rational mechanics. Why? Because

---

[6]There are three standard (or canonical) forms for elliptic integrals, introduced by Legendre in 1811.

elliptic functions were part of the undergraduate curriculum in Britain and the Continent until well into the twentieth century.

From the symmetries of the energy equation (56), it follows that the period T—the time required for the bob to regain its original position—is quadruple the time for the bob to fall from rest from its original position ($\vartheta = \alpha$) to the down position ($\vartheta = 0$). Thus,

$$(63) \qquad T = 4\sqrt{l/g} \int_0^{\pi/2} \frac{d\varphi}{\sqrt{1 - \sin^2 \tfrac{1}{2}\alpha \sin^2 \varphi}}$$

$$= 4\sqrt{l/g}\, K(\sin \tfrac{1}{2}\alpha).$$

The period is a nonconstant function of α. If α is small, then the integrand can be expanded as a binomial series, and the resulting integrand can be integrated term by term. The integrated series converges rapidly if α is small:

$$(64) \qquad T = 2\pi\sqrt{l/g}\,(1 + \tfrac{1}{4}\sin^2 \tfrac{1}{2}\alpha + \tfrac{9}{64}\sin^4 \tfrac{1}{2}\alpha + \cdots).$$

In the limit as $\alpha \to 0$, $T \to 2\pi\sqrt{l/g}$.

It is important to notice that this limiting result could have been obtained from the equation of motion

$$(65) \qquad \ddot{\vartheta} + \frac{g}{l}\sin\vartheta = 0$$

by supposing ϑ to be restricted to values so small that the approximation $\sin\vartheta \cong \vartheta$ is valid. The equation of motion becomes

$$(66) \qquad \ddot{\vartheta} + \frac{g}{l}\vartheta = 0,$$

the differential equation of simple harmonic motion. It is well known that every solution of this differential equation has period $2\pi\sqrt{l/g}$.

Observing the period of a pendulum of known length was until recently the most accurate way of determining g, the acceleration of gravity.[7] A pendulum whose length is such to make $T = 2$ is called a *seconds pendulum*. A seconds pendulum will have $l = g/\pi^2$.

---

[7] A few laboratories scattered throughout the world are equipped to make determinations of g by specialized and more accurate techniques. See Chester H. Page and Paul Vigoureux, *The International Bureau of Weights and Measures 1875–1975*, United States Government Printing Office, Washington, 1975, pp. 103–115.

EXERCISE I.7. A pendulum at the surface of the Earth is adjusted by comparison with an atomic clock until it is a seconds pendulum. It is then remembered that the gravitational field at the surface of the Earth is not uniform but is everywhere pointing toward the center of the Earth. Will this pendulum beat faster or slower than a true seconds pendulum (with respect to an assumed uniform gravitational field at the same place) and by how many seconds will they differ by the end of a year? [You will need to make some simple geometric approximations. Obtain values for the necessary physical quantities from a handbook.]

EXERCISE I.8. A particle is constrained to move on the curve $y = f(x)$ under the action of a constant field of force in the direction of the negative y-axis while the plane of the curve rotates with constant angular velocity around the y-axis. If the motion has a constant velocity, show that the curve has the form of a parabola with its vertex downwards.

8. Foucault's pendulum

This is an experiment devised by the French scientist Foucault in 1851 to demonstrate without reference to astronomical observations that the Earth rotates. The experiment consists of suspending a simple pendulum and observing that the plane in which the pendulum swings appears to be rotating slowly. The true explanation of this phenomenon is that the Earth (with the observer) rotates under the pendulum. In the setup at many observatories and museums, the rotation of the plane is vividly shown by having the bob knock over small markers evenly spaced in a circle around the pendulum.

Referring to Figure I.6, let the point of suspension be O and let the bob be at P. Set $OP = r$, where $r = lu$, l is constant, and u is a unit vector. Let k be a unit vector pointing along CO, where C is the center of the Earth. We assume that l is very small compared to r_0, the radius of the Earth. Then we can suppose that CO and CP are parallel. The forces acting on the bob are two: the tension T in the string, acting along PO, and the gravitational force, acting along $-k$. Hence, the force on the bob is

(67) $$F = -Tu - mgk.$$

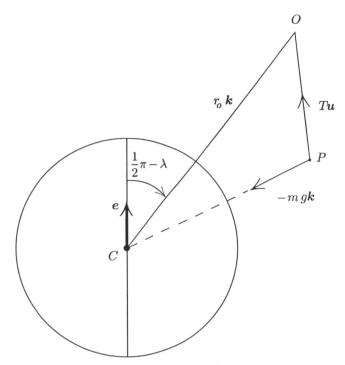

Figure I.6. Foucault's pendulum

The equation of motion as observed from O is obtained by using the expression (13) for the acceleration. The angular velocity ω of the Earth around its axis is so small that ω^2 can be neglected, so we drop the Coriolis acceleration term $\omega \times (\omega \times r)$. Because O itself is fixed to the Earth, the acceleration $a = 0$. Hence, if we use d/dt to refer to motion observed from O, then the equation of motion is

(68) $$\ddot{r} + 2\omega\, e \times \dot{r} = -gk - (T/m)u,$$

where e is a unit vector along the Earth's rotation axis, oriented so that the Earth's angular velocity is ωe. The usual equation for a simple pendulum drops out when $\omega = 0$.

To solve (68), eliminate T by forming the cross-product from the left with r. This gives

(69) $$r \times \ddot{r} + 2\omega r \times (e \times \dot{r}) = -g\, r \times k.$$

Now

(70)
$$r \cdot \dot{r} = \frac{1}{2}\frac{d}{dt}(r \cdot r) = \frac{1}{2}\frac{d}{dt}(l^2) = 0,$$

and $r \cdot e = -l \sin \lambda$, λ being the latitude of O. Thus, the triple product reduces to $-2\omega(r \cdot e)\dot{r} = -2l\omega \sin \lambda\, \dot{r}$. The equation of motion takes the form

(71)
$$r \times \ddot{r} - 2l\omega \sin \lambda\, \dot{r} = -glu \times k.$$

It is observed that the plane of the pendulum rotates around k. Take axes rotating with angular velocity vector $\omega' k$, and (recalling (12) and (13)) let $\partial/\partial t$ represent rate of change with respect to these axes. Because ω' is constant and is, we will assume, so small that $\omega'\omega$ and ω'^2 can be neglected,

(72)
$$r \times \left(\frac{\partial^2 r}{\partial t^2} + 2\omega' k \times \frac{\partial r}{\partial t}\right) - 2l\omega \sin \lambda \frac{\partial r}{\partial t} = -glu \times k.$$

(Here $\partial r/\partial t$ is written for dr/dt because $\omega'\omega$ is negligible.) Expanding the triple cross-product, we get

(73)
$$r \times \frac{\partial^2 r}{\partial t^2} - 2(r \cdot k\, \omega' + \omega l \sin \lambda)\frac{\partial r}{\partial t} = -glu \times k.$$

We assume that during the motion the pendulum is never far from vertical, so $u \cdot k \cong 1$. If $\omega' = -\omega \sin \lambda$, the differential equation reduces to

(74)
$$r \times \frac{\partial^2 r}{\partial t^2} = -gr \times k,$$

which is just the angular momentum-torque relation for a simple pendulum. Hence, the motion observed from O is that of a simple pendulum rotating around the vertical with angular velocity $-\omega \sin \lambda\, k$.

EXERCISE I.9. In your present location, how long does it take for the plane of a Foucault pendulum to make one full revolution? Through what angle does it turn in an hour?

II

Central
Forces

1. Motion in a central force field

A central force field is one whose action is always directed toward a fixed point. If that fixed point is taken to be the origin, and if i is taken to be a unit vector directed from the origin to the position r of a particle acted upon by the central force, then the force field can be written

$$(83) \qquad \boldsymbol{F}(\boldsymbol{r}) = -mP(\boldsymbol{r})\boldsymbol{i},$$

where P is a scalar function and m is the mass of the particle. The equation of motion for a particle of constant mass can be written down using Newton's Second Law; it is

$$(84) \qquad \ddot{\boldsymbol{r}} = -P\boldsymbol{i}.$$

Forming the cross-product on the left by r, we obtain

$$(85) \qquad \boldsymbol{r} \times \ddot{\boldsymbol{r}} = \boldsymbol{0}.$$

This differential equation can be integrated immediately to give

$$(86) \qquad \boldsymbol{r} \times \dot{\boldsymbol{r}} = \boldsymbol{h},$$

where $\boldsymbol{h}$, the *angular momentum per unit mass* (see I.2) is a constant vector. This is the *law of conservation of angular momentum.* Dotting the last equation with r, we find that

(79) $$r \cdot h = 0.$$

Therefore, the motion takes place in the plane perpendicular to h.

Under what condition is the force F derived from a potential energy function V? The radial and transverse components of ∇V are

(80) $$\frac{\partial V}{\partial r} \quad \text{and} \quad \frac{1}{r}\frac{\partial V}{\partial \vartheta},$$

r and ϑ now being polar coordinates in the plane containing the motion. The force being central, $\partial V/\partial \vartheta = 0$. Thus, the force is a function only of the radial distance: $P(\boldsymbol{r}) = P(r)$. But there is a potential energy function for *every* $P(r)$, namely $V(r) = m \int^r P(r)\, dr$. Therefore, a necessary and sufficient condition for a central force field to be conservative is that the magnitude of the force be a function only of the radial distance. In this case, the energy equation per unit mass is

(81) $$\tfrac{1}{2}|\dot{\boldsymbol{r}}|^2 + V(r) = \text{constant.}$$

From (17), we take the components of $\dot{\boldsymbol{r}}$ to write the energy equation (81) as

(82) $$\tfrac{1}{2}(\dot{r}^2 + r^2\dot{\vartheta}^2) + V(r) = \text{constant.}$$

If we write the equation of motion in terms of components, it becomes the pair of scalar equations

(83) $$\ddot{r} - r\dot{\vartheta}^2 = -P(r),$$
(84) $$r\ddot{\vartheta} + 2\dot{r}\dot{\vartheta} = 0.$$

The second equation can be multiplied by r, after which integration immediately gives

(85) $$r^2\dot{\vartheta} = h,$$

where h is a constant. By comparison with (78), it is easy to see that $h = |h|$.

The differential equation $r^2\dot\vartheta = h$ is expressed in terms of the areal velocity (20) in the form of *Kepler's Second Law,*

(86)
$$\frac{dA}{dt} = \tfrac{1}{2}h,$$

proposed by him for the motion of a planet in an elliptical orbit around the Sun but true for motion in any central force field:

For motion in a central force, the areal velocity is constant.

This often is described as the *law of conservation of angular momentum.*

To find the differential equation of the path described (often called the *orbit* or the *trajectory*), eliminate d/dt from the first polar equation (83) by using the relation

(87)
$$\frac{d}{dt} = \frac{h}{r^2}\frac{d}{d\vartheta}$$

derived from Kepler's Second Law. The result is

(88)
$$\frac{h}{r^2}\frac{d}{d\vartheta}\left(\frac{h}{r^2}\frac{dr}{d\vartheta}\right) - \frac{h^2}{r^3} = -P(r).$$

This differential equation is highly nonlinear, but a well-chosen substitution improves matters greatly. It is customary to let $u = 1/r$. After the usual manipulation, the last differential equation becomes

(89)
$$\frac{d^2u}{d\vartheta^2} + u = \frac{P(1/u)}{h^2u^2}.$$

While this differential equation is still nonlinear, the nonlinearity is 'tamer' than that in (88). The u-equation (89) is paired with Kepler's Second Law, which transforms into

(90)
$$\dot\vartheta = hu^2,$$

to give a pair of differential equations that should specify r and ϑ as functions of t.

Solution of the original system of two second-order differential equations for the radial and transverse accelerations will call for four constants. One of these, h, has already appeared when the transverse equation was integrated to yield Kepler's Second Law. The differential equation (89) will introduce two more. The fourth will enter during the determination of t by the equation

$$(91) \qquad t = \frac{1}{h} \int r^2 \, d\vartheta + \text{constant}.$$

(The original system, $\ddot{\boldsymbol{r}} = -P(\boldsymbol{r})\boldsymbol{i}$, is sixth order. The fifth and sixth constants are used to specify the direction of $\boldsymbol{h}$ in space.)

If we refer the orbit to Cartesian coordinates with respect to axes fixed in the plane of motion, say with $x = r \cos \vartheta$ and $y = r \sin \vartheta$, then the Kepler's Second Law—the conservation of angular momentum—is expressed as

$$(92) \qquad x\dot{y} - y\dot{x} = h.$$

EXERCISE II.1. A particle is attracted to the origin by a force proportional to its distance from the origin. (This is *Hooke's Law*.) Show that the motion takes place along an ellipse centered at the origin.

2. Force and orbit

We ask for the law of force that must act toward a given point in order for a given curve to be an orbit. To answer this question, it is convenient to introduce a different set of coordinates.

Suppose a point moves along a certain curve in the plane. Let r be the distance of that point from the origin, p the length of the perpendicular from the origin onto the tangent to the curve at the point, s an arc length variable, ρ the radius of curvature of the path at the point, and $v = \dot{s}$ the speed, all at time t. (See Figure II.1.) Set $h = pv$. Then we have *Sciacci's Theorem:*[1]

> The acceleration of the point can be resolved into components $\dfrac{h^2 r}{p^3 \rho}$ along the radius vector to the origin and $\dfrac{h}{p^2} \dfrac{dh}{ds}$ along the tangent to the path.

---

[1] Speakers of English who do not also speak Italian may pronounce 'Sciacci' as Shock'y.

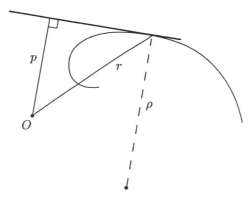

Figure II.1. Tangent line and radius of curvature

To see this, note that the acceleration can be resolved into components (see Figure II.2).

(93) $$\frac{dv}{dt} = v\frac{dv}{ds} \quad \text{along the tangent}$$

and

(94) $$\frac{v^2}{\rho} \quad \text{along the normal.}$$

But a vector F directed outward from the origin can be resolved into vectors $-|F|p/r$ directed along the inward normal and $|F|\,dr/ds$ along the tangent (this is geometrically clear!), so a vector v^2/ρ along the inward normal came from a force of magnitude $rv^2/\rho p$ directed inward along the radius vector having a component $\dfrac{rv^2}{\rho p}\dfrac{dr}{ds}$ along the tangent. The acceleration is therefore equivalent to components

(95) $$v\frac{dv}{ds} + \frac{rv^2}{\rho p}\frac{dr}{ds} \quad \text{along the tangent}$$

and

(96) $$\frac{rv^2}{\rho p} \quad \text{inwards along the radius.}$$

But $v = h/p$, so the latter component is $\dfrac{h^2 r}{p^3 \rho}$. We will show that $\dfrac{r}{\rho}\dfrac{dr}{ds} = \dfrac{dp}{ds}$

or $\dfrac{r}{\rho} = \dfrac{dp}{dr}$. In terms of the unit tangent and normal vectors, we have

$v = dr/ds = vt$, $p = -n \cdot r$, so

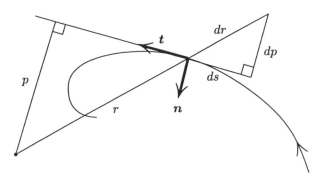

Figure II.2. Acceleration components

(97)
$$\frac{dp}{ds} = -\frac{dn}{ds} \cdot r - n \cdot \frac{dr}{ds} = \frac{r}{\rho} \sin \angle(p, r) = \frac{r}{\rho} \frac{dr}{ds},$$

as claimed. Thus, the former component of acceleration can be written

(98)
$$v \frac{dv}{ds} + \frac{rv^2}{\rho p} \frac{dr}{ds} = \frac{1}{2p^2} \frac{d(v^2 p^2)}{ds} = \frac{h}{p^2} \frac{dh}{ds},$$

which is Sciacci's Theorem!

Now we can describe the field of force which must act toward a given point (the origin) in order that a given curve may be described. If the curve is given in polar coordinates, the magnitude of the force is obtained from (89) as

(99)
$$P = h^2 u^2 \left(u + \frac{d^2 u}{d\vartheta^2} \right).$$

If the curve is given in (r, p) coordinates, the force is given by (97) and Sciacci's Theorem:

(100)
$$P = \frac{h^2}{p^3} \frac{dp}{dr}.$$

EXERCISE II.2. Find the acceleration of a point that describes a logarithmic spiral with constant angular velocity around the pole. Under what central force is a logarithmic spiral described? [You can use Sciacci's Theorem or work the force out directly.]

EXERCISE II.3. If the tangential and normal components of the acceleration of a point moving in the plane are constant, then the point describes a logarithmic spiral.

If the equation of the curve is given in rectangular coordinates, then the central force field is found in the following way.

Take the center of force to be the origin and let $f(x, y) = 0$ be the equation of the given curve. The equation of angular momentum is given by

$$(101) \qquad\qquad x\dot{y} - y\dot{x} = h.$$

From the equation of the curve,

$$(102) \qquad\qquad f_x\dot{x} + f_y\dot{y} = 0,$$

where $f_x = \partial f/\partial x$, etc. From these two equations, we calculate

$$(103) \qquad\qquad \dot{x} = \frac{-hf_y}{xf_x + yf_y} \quad\text{and}\quad \dot{y} = \frac{hf_x}{xf_x + yf_y}.$$

Another differentiation gives

$$
\begin{aligned}
\ddot{x} &= \dot{x}\frac{\partial \dot{x}}{\partial x} + \dot{y}\frac{\partial \dot{x}}{\partial y} \\
(104) &= \frac{-hf_y}{xf_x + yf_y}\frac{\partial}{\partial x}\left(\frac{-hf_y}{xf_x + yf_y}\right) + \frac{hf_x}{xf_x + yf_y}\frac{\partial}{\partial y}\left(\frac{-hf_y}{xf_x + yf_y}\right) \\
&= \frac{h^2 x(-f_y^2 f_{xx} + 2f_x f_y f_{xy} - f_x^2 f_{yy})}{(xf_x + yf_y)^3}.
\end{aligned}
$$

However, the required force is P and $\ddot{x} = -Px/r$, so that the required central force is

$$(105) \qquad\qquad P = \frac{h^2 r(f_y^2 f_{xx} - 2f_x f_y f_{xy} + f_x^2 f_{yy})}{(xf_x + yf_y)^3}.$$

EXAMPLE. Take a conic in the form

$$(106) \qquad 2f(x, y) \equiv ax^2 + 2hxy + by^2 + 2gx + 2fy + c = 0.$$

Then the quantity $f_y^2 f_{xx} - 2f_x f_y f_{xy} + f_x^2 f_{yy}$ has the constant value $-(abc + 2fgh - af^2 - bg^2 - ch^2)$, while the quantity $xf_x + yf_y$ has the value $-(gx + fy + c)$.

Associated to the conic and a point (x_1, y_1) of the plane is the line with the equation

(107)
$$(ax_1 + hy_1 + g)x + (hx_1 + by_1 + f)y + (gx_1 + fy_1 + c) = 0.$$

This line is the *polar* of the point with respect to the line.[2] Therefore, $-(gx + fy + c)$ is a constant multiple[3] of the perpendicular distance from the point (x, y) on the conic onto the polar of the origin with respect to the conic.

We have arrived at a result due to Hamilton:

> A given conic can be described by a central force acting on a particle in the position (x, y) which varies directly as the radius from the center of the force to (x, y) and inversely as the cube of the perpendicular from (x, y) onto the polar of the center of force.

3. The integrable cases of central forces

We consider central forces whose magnitude depends only upon the distance r. In terms of $u = 1/r$, the differential equation of the orbit is (89):

(108)
$$\frac{d^2u}{d\vartheta^2} + u = \frac{P(1/u)}{h^2 u^2}.$$

Multiply by $2\,du/d\vartheta$ and integrate, leading to

(109)
$$\left(\frac{du}{d\vartheta}\right)^2 = C - \frac{2}{h^2} \int^u \frac{P(1/u)}{u^2}\,du - u^2,$$

where C is a constant. This equation is separable. Restoring the variable r and separating, we get

(110)
$$\vartheta = \int^r \left\{ C - \frac{2}{h^2} \int^r P(r)\,dr - \frac{1}{r^2} \right\}^{-1/2} \frac{dr}{r^2}.$$

---

[2] The concept and the name come from duality theory in projective geometry. If (x_1, y_1) lies on the conic, then the polar becomes the tangent line to the conic at that point.

[3] The multiplier is $-\sqrt{f^2 + g^2}$.

This is the equation of the orbit in polar coordinates. In principle, we perform the integrations, solve the resulting relation for r as a function of ϑ, and determine the time from the conservation of angular momentum by the integral

$$(111) \qquad t = \frac{1}{h} \int^\vartheta r^2\, d\vartheta + \text{constant}.$$

The process of integration was once commonly called *quadrature,* referring to its interpretation as finding an area. Therefore, we have shown that the problem of motion under central forces is always solvable by quadratures when the force is a function of distance only.

It remains to consider the cases when the quadratures can be carried through in terms of 'known' functions.

Suppose that the force varies as the nth power of the distance, so as u^{-n}. The integral for (110) for the determination of ϑ can be written in the form

$$(112) \qquad \vartheta = \int (a + bu^2 + cu^{-n-1})^{-1/2}\, du,$$

where a, b, c are constants, with the exception of $n = -1$ when a logarithm replaces the power u^{-n-1}. If the integral is to be expressible in terms of *circular* functions, then the integrand can involve the square root of a polynomial of degree at most 2. This gives $-n - 1 = 0, 1, 2$, so $n = -1, -2, -3$. However, $n = -1$ has already been excluded. Furthermore, $n = 1$ must be added, for then the irrationality becomes quadratic when u^2 is taken as the new variable.[4]

Many other exponents lead to integrations in terms of elliptic functions. For this to be possible, it is necessary for the radicand to be a polynomial of degree 3 or 4, or to be transformable to one by introduction of a new variable of integration. The suitable exponents are

$$(113) \qquad n = 5, 3, 0, -\tfrac{1}{3}, -\tfrac{3}{2}, -\tfrac{5}{3}, -\tfrac{7}{3}, -\tfrac{5}{2}, -4, -5, -7.$$

EXAMPLE. We examine motion under the inverse cube law of force.

---

[4] We have already discussed this case in §1.

The attractive force is $P = \mu/r^3$, so the polar equation of the orbit becomes

(114)
$$\vartheta = -\int^u \left\{ C + \left(\frac{\mu}{h^2} - 1 \right) u^2 \right\}^{-1/2} du.$$

Generically, $C \neq 0$. Setting $k^2 = |1 - \mu/h^2|$, integration leads in the generic case to

(115)
$$u = \begin{cases} A\cos(k\vartheta + \epsilon), & \text{if } \mu < h^2, \\ A\vartheta + \epsilon, & \text{if } \mu = h^2, \\ A\cosh(k\vartheta + \epsilon) & \text{if } \mu > h^2. \end{cases}$$

In each case, A and ϵ are constants of integration. The first and third cases are called *Cotes's spirals*, and the second case is the *reciprocal spiral*.

EXERCISE II.4. Derive the Cotes's spirals by beginning with the differential equation (89). Identify the orbits in the nongeneric cases, when $C = 0$.

It can be shown that the only central forces for which every bounded orbit is closed (that is, periodic) are those given by the power laws r^{-2} and r^1. This result is called *Bertrand's Theorem*, and it will be proved in Chapter V.

EXERCISE II.5. Sketch a sample solution curve in the (r, ϑ) coordinate system for each of the three cases of the inverse cube law.

EXERCISE II.6. If an orbit $r = f(\vartheta)$ be described under the central force $P(r)$ to the origin, then the orbit $r = f(k\vartheta)$, where k is any constant, can be described under a central force $k^2 P(r) + c/r^3$, where c is a suitable constant. Furthermore, the intervals of time between corresponding points are the same on the two orbits.

An *apse* of an orbit is a point where the radius r takes a maximum or a minimum. At an apse, $dr/d\vartheta = 0$.

EXERCISE II.7. A particle of mass m is projected from an apse under the attraction of a force

(116)
$$\frac{m}{r^3} \log \frac{r}{\alpha}$$

directed to a center at distance α from the apse. The initial velocity is that which the particle would have were it to fall to the apsidal point from a rest position at infinity. Find the equation of the orbit described.

EXERCISE II.8. A particle is subject to a central attraction of magnitude μ/r^3 toward a fixed point O. It is projected from a point P with speed $\sqrt{\mu}/OP$, in a direction making an angle α with the radius vector OP. Prove that the tangent to the path at any point makes a fixed angle with the radius vector, and name the trajectory.

EXERCISE II.9. A particle moves under an attraction

(117) $$\frac{\mu}{r^2} + \frac{\nu}{r^3}$$

to a fixed center. Show that the angle subtended at the center of force by two consecutive apses is

(118) $$\frac{\pi}{\sqrt{1 - \nu/h^2}},$$

where h is the constant of angular momentum.

4. Bonnet's Theorem

Despite its name, this theorem was discovered by Legendre and published by him in 1817. Here is *Bonnet's Theorem:*

> If a given orbit can be described in each of n given fields of force taken separately, the velocities at any point P of the orbit being $v_1, v_2, \ldots, v_n$, respectively, then the same orbit can be described in the field of force which is obtained by superimposing all of these fields, the velocity at the point P being
> $$(v_1^2 + v_2^2 + \cdots + v_n^2)^{1/2}.$$

To prove Bonnet's Theorem, suppose that with the field of force that is the resultant of superimposing the original fields an additional force of magnitude R normal to the curve is required to make the point move along the curve. Suppose the particle to be projected from a point A on the orbit in such a way that the square of its velocity at A is the sum of the squares of its velocities at A in the original fields of force. If v is the velocity at P under the action of the combined forces while v_i is that produced by the ith original force, then the respective kinetic energies are $\frac{1}{2}mv^2$ and $\frac{1}{2}mv_i^2$. Let s be an arc-length variable along the orbit, t the (unit) tangent vector, and a and a_i the accelerations produced by the resultant force F and by the ith original force F_i. Then

(119)
$$\frac{d}{ds}(\tfrac{1}{2}mv^2) = mv\cdot\frac{dv}{ds} = mt\cdot\frac{dv}{dt} = t\cdot F$$
$$= t\cdot\sum F_i = \sum(t\cdot F_i)$$
$$= \frac{d}{ds}\sum(\tfrac{1}{2}mv_i^2).$$

Therefore, v^2 and $(v_1^2 + v_2^2 + \cdots + v_n^2)^{1/2}$ differ by a constant for all time. Because they were made equal at A, they are equal at any P on the orbit.

Let ρ be the radius of curvature of the orbit. The normal component of the force on the particle is

(120)
$$\frac{mv^2}{\rho} = \frac{m(v_1^2 + v_2^2 + \cdots + v_n^2)^{1/2}}{\rho} = F^\perp + R,$$

where $F^\perp$ is the component of F normal to the orbit. But

(121)
$$\frac{mv_i^2}{\rho} = F_i^\perp$$

and $F^\perp = \sum F_i^\perp$. Therefore, $R = 0$, and Bonnet's Theorem follows.

EXERCISE II.10. After hearing the statement of Bonnet's Theorem, a student asked what would would happen if there were just two forces, F_1 and F_2, such that $F_1 + F_2 = 0$. How would you answer that question?

5. Miscellaneous exercises

Reader, here is an opportunity for you to pause during your career through the celestial spheres and refresh yourself by working out a few exercises.

EXERCISE II.11. A particle moves in a smooth, circular tube under the influence of a force directed to a fixed point in its plane and proportional to the distance from that point. Show that the motion has the same character as the motion of a simple pendulum (I.7).

[Heuristically, one can imagine the center of force withdrawing to infinity. When the center is at a very great distance, the force vectors on the circle will be essentially parallel and the magnitude of the force will be essentially constant across the circle. Thus, the motion to be examined here has the simple pendulum motion as a limiting case.]

Show as a consequence that a particle which is constrained to move under no external forces in a plane circular tube which itself is constrained to rotate uniformly about any point in its plane moves in a way similar to the motion of a simple pendulum.

EXERCISE II.12. A particle moves on a straight line under the action of two centers of repulsive force of equal strength μ^2, the magnitude of the repulsion varying as the inverse square of the distance. Show that if the centers of the force are at distance $2c$ apart and the particle starts from rest at a distance kc ($k < 1$) from the middle point of the line joining them, then it will perform oscillations of period

(122)
$$\frac{2\sqrt{c^3(1 - k^2)}}{\mu} \int_0^{\pi/2} (1 - k^2 \sin^2 \vartheta)^{1/2} \, d\vartheta.$$

[We have here the *complete elliptic integral of the second kind* (cf. (60)), for which the standard notation is

(123)
$$E(k) = \int_0^{\pi/2} (1 - k^2 \sin^2 \vartheta)^{1/2} \, d\vartheta.$$

This integral is encountered in the rectification of the ellipse.]

EXERCISE II.13. Show that the force perpendicular to the asymptote under which the curve $x^3 + y^3 = a^3$ can be described is proportional to $xy(x^2 + y^2)^{-3}$.

EXERCISE II.14. If a circle is described under a central attraction directed to a point on its circumference, then the law of force is the inverse fifth power of the distance.

[In his investigations of the dynamical theory of gases, Maxwell assumed that the molecules of gas mutually interact according to the inverse fifth power law. A glance at the list (113) of exponents whose orbits can be described in terms of elliptic functions shows that the inverse fifth power law is one of them. Maxwell admitted that his choice of the exponent -5 came less from the physical consequences of the choice than from the attractiveness of the possibility of explicit integration of the equations of motion.]

EXERCISE II.15. A particle of unit mass describes an orbit under an attractive force P to the origin and a transverse force T perpendicular to the radius vector. Prove that the differential equation of the orbit is given by

(124)
$$\frac{d^2u}{d\vartheta^2} + u = \frac{P}{h^2 u^2} - \frac{T}{h^2 u^3} \frac{du}{d\vartheta} \quad \text{and} \quad \frac{d(h^2)}{d\vartheta} = \frac{2T}{u^3}.$$

If the attractive force is always zero and if the particle moves on a logarithmic spiral with constant angle α between the radius vector and the tangent line, show then that

(125)
$$T = \mu r^{2 \sec^2 \alpha - 3} \quad \text{and} \quad h = (\mu \sin \alpha \cos \alpha)^{1/2} r^{\sec^2 \alpha}.$$

EXERCISE II.16. A particle moves on the surface of a cone under a force directed to the vertex and depending only upon the distance to the vertex. Show that the orbit is a plane section of the cone if and only if the magnitude of the force is

(126) $$\frac{A}{r^2} - \frac{B}{r^3},$$

where r is the distance to the vertex. [Suggestion: Unroll the cone into the plane and transform the problem into one of central forces in the plane.]

6. Motion on a surface of revolution

One important case of motion on a surface which is solvable by quadratures is that of a particle which moves on a surface of revolution under forces derivable from a potential energy function which is symmetrical with respect to revolution around the same axis in space.

Define the position of a point in space by cylindrical coordinates (r, ϑ, z), where z is a coordinate measured parallel to the axis of the surface, r is the perpendicular distance of the point from this axis (not to be confused with the earlier use of r to denote distance to a center of force), and ϑ is the azimuthal angle made by r with a fixed plane through the axis. The shape of the surface will be described by an equation relating r and z, say

(127) $$r = f(z).$$

The potential energy can not involve ϑ; therefore, it is a function only of r and z. On the surface, r is determined by f as a function of z, so that the potential energy can be expressed for points on the surface as a function only of z, say as $V(z)$. For simplicity, we take the mass of the particle to be unity, thereby identifying the potential energy function with the same-valued potential function.

Let k be a unit vector along the axis of revolution in the direction of increasing z; let i point from the axis of revolution, perpendicular to it and through the particle; and let j complete the positively oriented orthonormal triple i, j, k. Denote the position vector with respect to the origin by r. Because of the rotational symmetry, there are scalar functions G and H of z such that the equation of motion can be written as

(128) $$\ddot{r} = Gi + Hk.$$

Form the cross-product on the left with $r = ri + zk$ and dot the result on the right by k:

(129) $$r \times \ddot{r} \cdot k = (ri + zk) \times (Gi + Hk) \cdot k = 0.$$

This equation integrates immediately to give

(130) $$r \times \dot{r} \cdot k = \text{constant},$$

which can be written as

(131) $$r^2 \dot{\vartheta} = h,$$

in which h is a constant. We have obtained an equation that plays the role of an angular momentum integral of the motion.

Because there is a potential energy function,[5] a second integral, expressing conservation of energy, can be written down immediately:

(132) $$\tfrac{1}{2}(\dot{r}^2 + r^2\dot{\vartheta}^2 + \dot{z}^2) + V(z) = \text{constant}.$$

We have assumed a relation $r = f(z)$. Suppose further that E is the total energy per unit mass. Then

(133) $$\tfrac{1}{2}\{[f'(z)^2 + 1]\dot{z}^2 + f(z)^2\dot{\vartheta}^2\} + V(z) = E.$$

The 'momentum' equation (131) gives $\dot{\vartheta} = h/f(z)^2$, so that

(134) $$[f'(z)^2 + 1]\dot{z}^2 + \frac{h^2}{f(z)^2} + 2V(z) = 2E.$$

This equation is clearly integrable by quadratures, the result being

(135)
$$t + \text{constant} = \int [f'(z)^2 + 1]^{1/2}[2E - 2V(z) - h^2/f(z)^2]^{-1/2}\, dz.$$

The values of r and ϑ are then obtained from the equation $r = f(z)$ of the surface and from solution by quadrature of the 'momentum' equation, $f(z)^2\dot{\vartheta} = h$, respectively.

---

[5] Recall that we have supposed the particle to have unit mass.

EXAMPLE. A particle moves on the vertical circular cylinder $r = a$ under the downward attraction of gravity. We take $V(z) = z$. The integral (135) for time becomes

$$(136) \qquad t + \text{constant} = \int [2E - 2gz - h^2/a^2]^{-1/2} \, dz.$$

Add a suitable constant to the potential energy V, if necessary, to ensure that $2Ea^2 = h^2$. Then

$$(137) \qquad t + \text{constant} = \int (-2gz)^{-1/2} \, dz,$$

so that

$$(138) \qquad z = -\tfrac{1}{2}g(t - t_0)^2.$$

The equation of angular momentum is $a^2 \dot{\vartheta} = h$, which yields

$$(139) \qquad \vartheta - \vartheta_0 = \frac{E}{a^2}(t - t_0).$$

Both t_0 and ϑ_0 are constants.

The particle moves vertically as if it were falling from rest, while it turns around the axis of the cylinder at a constant rate.

Motion on the sphere, paraboloid, and cone under gravity directed downward along the axis of rotation can be expressed in terms of elliptic functions.

EXERCISE II.17. By developing the cone into a plane (or otherwise), show that motion on a conical surface under gravity directed downward along the axis is one of the cases of motion integrable in terms of elliptic functions. [Consult the list (113).]

EXERCISE II.18. A heavy particle is projected horizontally with a velocity v inside a smooth sphere at an angular distance α from the vertical radius drawn downward. Show that it will never fall below nor ever rise above its initial level according to whether

$$(140) \qquad v > ag \sin \alpha \tan \alpha, \quad \text{or} \quad v < ag \sin \alpha \tan \alpha,$$

where a is the radius of the sphere.

Show furthermore that the highest point attained on the spherical surface is at an angular distance β from the lowest point attained, where β is the smaller of the values λ, μ given respectively by the equations

$$(141) \qquad (3 \cos \lambda - 2 \cos \alpha)ag + v^2 = 0,$$

$$(142) \qquad (\cos \mu + \cos \alpha)v^2 - 2ag \sin^2 \mu = 0.$$

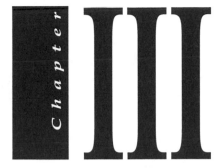

Orbits under the Inverse Square Law

1. Kepler's three laws and Newton's Law

Kepler formulated his famous three laws of planetary motion during the first decades of the seventeenth century. It was an arduous task, proceeding strictly from observed data, the most accurate of which were contributed by Kepler's mentor, Tycho Brahe.

> *Kepler's First Law:*   The orbit of a planet around the Sun
> is an ellipse, with the Sun being situated at a focus.

Analytically, if the Sun is at the focus S and P is the instantaneous position of the planet, the polar equation of the ellipse is

$$(151) \qquad r = \frac{a(1 - e^2)}{1 + e\cos(\vartheta - \omega)},$$

where r is the *radius vector* SP; ϑ is the oriented angle in the plane of the ellipse measured in the direction of the planet's motion to SP from a suitably chosen reference ray SN; ω is the angle from SN to SA, where A is the point of the ellipse closest to S; a is the *semi-major axis*; and e is the *eccentricity* of the ellipse. In Figure III.1, C is the center of the ellipse, $CA = CB = a$, and $CS = ae$. (Thus, $0 \le e < 1$.)

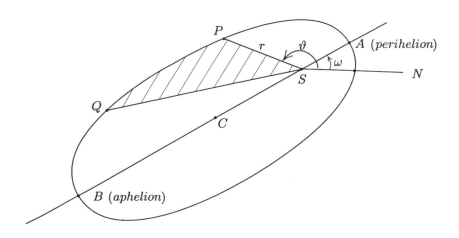

Figure III.1. Elliptical orbit

From the polar equation (143), the minimum value of r is $a(1 - e)$, corresponding to the point A at which $\vartheta = \omega$; A is called the *perihelion,* and it is an apse.[1] The maximum value of r is $a(1 + e)$, corresponding to the point B with $\vartheta = \omega \pm \pi$; it is again an apse, called the *aphelion.*

> *Kepler's Second Law:* The rate of description of area by the moving radius vector is constant in time.

Thus, if A represents the cumulative area swept out in time by the planet, then $dA/dt = $ constant. But $dA/dt = \frac{1}{2}r^2\dot{\vartheta}$, so the Second Law can be written as

$$(144) \qquad\qquad r^2\dot{\vartheta} = h,$$

where h is a constant equal to twice the rate of sweeping out area. Clearly, the Second Law expresses the law of conservation of angular momentum and, as such, is true for central forces whose magnitude depends only upon the radial distance.

---

[1] See p. 34.

Denote by T the time required for the planet to describe a complete orbit; this is the (*orbital*) *period*. Because the area of the ellipse is πab, where $b = a\sqrt{1-e^2}$ is the *semi-minor axis*, the constant h is by the Second Law equal to $2\pi ab/T$, so that

$$(145) \qquad h = \frac{2\pi}{T} a^2\sqrt{1-e^2}.$$

In time T the radius sweeps out an angle 2π; the *mean angular motion* n is defined as

$$(146) \qquad n = \frac{2\pi}{T}.$$

Then we can write

$$(147) \qquad h = na^2\sqrt{1-e^2}.$$

Newton used Kepler's First and Second Laws to obtain an expression for the magnitude of a central force required a particle in a planetary orbit. We found in §2 expressions for such forces in terms of the equation of the orbit. In particular (89), the force has magnitude

$$(148) \qquad P = h^2 u^2 \left(u + \frac{d^2 u}{d\vartheta^2} \right).$$

Here,

$$(149) \qquad u = \frac{1 + e\cos(\vartheta - \omega)}{a(1 - e^2)},$$

and an easy calculation shows that the magnitude of the force is

$$(150) \qquad \frac{h^2}{a(1 - e^2)} \frac{1}{r^2}.$$

Newton's *Law of Universal Gravitation* states that a particle of mass m_1 attracts a particle of mass m_2 with a force of magnitude F, acting along the line joining the particles, which is proportional to the product of the masses and inversely proportional to the square of the distance, r, between the particles. Expressed algebraically, the law is

$$(151) \qquad F = \frac{Gm_1 m_2}{r^2},$$

where G is called the *constant of gravitation*.

This expression also gives the magnitude of the attraction of m_1 by m_2. Furthermore, the acceleration f_1 of m_1 due to the force F is given by $f_1 = Gm_1/r^2$.

The constant G is very difficult to determine with high precision. The accepted value is

(152) $$G = 6.66 \times 10^{-8} \, \text{cm}^3 \, \text{g}^{-1} \, \text{sec}^{-2}.$$

The last of Kepler's Laws is

> *Kepler's Third Law:* If planets $P_1, P_2, \ldots$ revolve around the Sun, then
>
> (153) $$\frac{a_1^3}{T_1^2} = \frac{a_2^3}{T_2^2} = \cdots = \text{constant.}$$

Alternatively, using (146),

(154) $$n_1^2 a_1^3 = n_2^2 a_2^3 = \cdots = \text{constant.}$$

We will derive Kepler's Third Law from Newton's Law of Universal Gravitation.

2. The orbit from Newton's Law

Let the Sun have mass m_0 and the planet mass m. The two bodies, *supposed to be particles,*[2] attract each other with a force directed along the line joining them and of magnitude

(155) $$\frac{Gmm_0}{r^2},$$

where r is the distance between the particles. To avoid certain considerations about centers of mass, we will make the problem a little abstract.

Suppose that S is the center of attractive force of magnitude μ/r^2, and suppose the planet to have mass m. Then the motion is planar and the equations of motion in polar coordinates centered at S have been derived earlier (83), (84); they are

(156) $$\ddot{r} - r\dot{\vartheta}^2 = -P(r),$$

(157) $$r\ddot{\vartheta} + 2\dot{r}\dot{\vartheta} = 0.$$

---

[2]We will examine this assumption on page 126.

We have already seen that the second of these equations is the analytical expression of Kepler's Second Law, h being twice the rate of description of area by the radius vector.

To obtain the polar equation of the orbit, eliminate t between the two differential equations and then write $u = 1/r$. Because $\dot{\vartheta} = hu^2$, we have

$$(158) \qquad \dot{r} = \frac{d(1/u)}{d\vartheta}\,\dot{\vartheta},$$

and

$$(159) \qquad \ddot{r} = \frac{d}{d\vartheta}\left(-h\frac{du}{d\vartheta}\right)\dot{\vartheta} = -h^2 u^2 \frac{d^2 u}{d\vartheta^2}.$$

The first equation of motion becomes

$$(160) \qquad \frac{d^2 u}{d\vartheta^2} + u = \frac{\mu}{h^2}.$$

The right-hand side of the differential equation (160) is a constant; the general solution is the sum of a periodic solution to the homogeneous equation and a constant. It is easy to obtain the solution in the form

$$(161) \qquad u = \frac{\mu}{h^2}\left\{1 + e\cos(\vartheta - \omega)\right\},$$

in which e and ω are constants of integration. Hence

$$(162) \qquad r = \frac{h^2/\mu}{1 + e\cos(\vartheta - \omega)}.$$

This is the polar equation of a conic section: a hyperbola if $e > 1$ and a parabola if $e = 1$. If $0 \le e < 1$, it is an ellipse of eccentricity e. Excluding $e = 0$, which is the degenerate case of a circle, the semi-latus rectum is h^2/μ. Introduce a, the *semi-major axis,* satisfying

$$(163) \qquad h^2 = \mu a(1 - e^2),$$

and n, the *mean angular motion,* satisfying

$$(164) \qquad h = na^2\sqrt{1 - e^2}.$$

Then simple algebra gives

$$(165) \qquad \mu = n^2 a^3.$$

In Newton's Law of Gravitation, $\mu = Gm_0$; furthermore, the period $T = 2\pi/n$. Therefore,

$$(166) \qquad \frac{a^3}{T^2} = \frac{Gm_0}{4\pi^2},$$

which is Kepler's Third Law.

We will be concerned with elliptical orbits. Such orbits are specified by six constants, called *elements* of the orbit. These are traditionally taken to be

$$(167) \qquad a, e, \tau, \Omega, \omega, i.$$

Here, a, e, ω have the meanings already explained; τ is the time of perihelion passage; and Ω, i serve to position the plane of the orbit with respect to the celestial sphere. We however will ignore Ω, i and concentrate upon the elements a, e, τ that refer to the ellipse regardless of its orientation in the reference plane and in space.

EXERCISE III.1. Solve the simultaneous differential equations

$$(168) \qquad \frac{d^2r}{dt^2} - r\left(\frac{d\vartheta}{dt}\right)^2 = 0 \qquad and \qquad r\frac{d^2\vartheta}{dt^2} + 2\frac{dr}{dt}\frac{d\vartheta}{dt} = 0.$$

[You could write a single vector equation which encompasses the given two equations and whose solution is seen without effort to represent motion along a straight line. In the spirit of this section, eliminate t between the two differential equations and show that the resulting r, ϑ differential equation can be solved to give motion along a straight line.]

3. The true, eccentric, and mean anomalies

The *true anomaly,* historically denoted by f, is defined to be the oriented angle $\angle AP_0P$, where now the focus of the ellipse is denoted by P_0. (See Figure III.2.) In the notation of the preceding section, $f = \vartheta - \omega$. Then the equation of the orbit is

$$(169) \qquad r = \frac{a(1 - e^2)}{1 + e\cos f}.$$

Evidently, the equation of areas (Kepler's Second Law) is

$$(170) \qquad r^2\dot{f} = h.$$

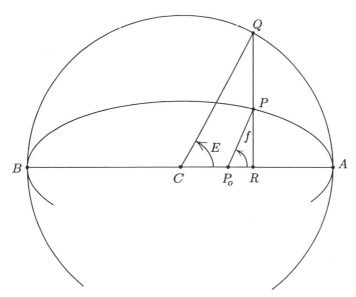

Figure III.2. True and eccentric anomalies

The polar coordinates of P with respect to P_0A as initial ray are (r, f). The corresponding rectangular coordinates are P_0R and RP. Denoting these by (ξ, η), we have

(171) $$\xi = r \cos f \quad \text{and} \quad \eta = r \sin f.$$

Draw a circle with C as center and radius a. The ordinate of P cuts this circle at Q. Historically, the angle $\angle ACQ$ is called the *eccentric anomaly* and it is denoted by E. Then

(172) $$CR = a \cos E \quad \text{and} \quad RQ = a \sin E.$$

Because the equation of the ellipse, referred to C as origin and CA as positive x-axis, is

(173) $$\frac{x^2}{a^2} + \frac{y^2}{b^2} = 1,$$

where $b = a\sqrt{1 - e^2}$, the ordinate RP is given in terms of E by

(174) $$RP = \frac{b}{a}\sqrt{a^2 - a^2 \cos^2 E} = b \sin E.$$

Hence, $RP : RQ = b : a$. Because $CP_0 = ae$, we have

(175) $\quad \xi \equiv r \cos f = a(\cos E - e) \quad$ and $\quad \eta \equiv r \sin f = b \sin E$.

The radius vector is found by eliminating f between the coordinate equations. Because $b^2 = a^2(1 - e^2)$, we find

(176) $$r = a(1 - e \cos E).$$

Using a double-angle formula on the ξ-equation of (175) gives

(177) $$r \left(2 \cos^2 \tfrac{1}{2} f - 1 \right) = a(\cos E - e).$$

Combining this relation with the expression of r in terms of E, we get

(178) $$\begin{aligned} r \cos^2 \tfrac{1}{2} f &= \tfrac{1}{2} a[1 - e \cos E + \cos E - e] \\ &= \tfrac{1}{2} a(1 - e)(1 + \cos E) \\ &= a(1 - e) \cos^2 \tfrac{1}{2} E. \end{aligned}$$

Similarly,

(179) $$r \left(1 - 2 \sin^2 \tfrac{1}{2} f \right) = a(\cos E - e),$$

leading to

(180) $$r \sin^2 \tfrac{1}{2} f = a(1 + e) \sin^2 \tfrac{1}{2} E.$$

By division, we get an equation relating the true anomaly and the eccentric anomaly:

(181) $$\tan \tfrac{1}{2} f = \left(\frac{1 + e}{1 - e} \right)^{1/2} \tan \tfrac{1}{2} E.$$

We suppose that the planet is at point P at time t. In the interval from τ to t, the radius vector has swept out the sector AP_0P. The angle swept out in the same time by a radius vector moving with the constant mean velocity $n = 2\pi/T$, the position of the radius vector at time τ being along P_0A, is called the *mean anomaly*. If we denote the mean anomaly by M, then

(182) $$M = n(t - \tau).$$

To illustrate the three anomalies, consider the libration in longitude of the Moon. The period of rotation of the Moon on its axis is the same as the period of its revolution around the Earth (one *sidereal month*). If the Moon revolved around the Earth in a circular orbit, then the same half of its surface would always point toward the Earth, and we would see nothing of the other half. But the Moon moves in an elliptical orbit with the Earth at P_0, a focus. Let the center of the Moon be at P, and let S be that point on the surface of the Moon such that P_0, S, and P are collinear at perigee.[3] At any time, PS will make an angle with the major axis CA of the ellipse that is equal to the mean anomaly of the Moon in its orbit, while the angle between PP_0 and P_0A is the true anomaly f. Assuming for the while that the axis of rotation of the Moon is perpendicular to the plane of its orbit, an observer on the Earth will see S as being displaced from the center of the Moon's disc by a longitude $M - f$, or he will see this extra longitude at the Moon's limb compared to what he saw at perigee. This is known as the *libration in longitude* of the Moon.[4] It will reach its maximum value when

(183)
$$\frac{d(M - f)}{dt} = 0,$$

or

(184)
$$\frac{df}{dt} = n.$$

By the equation of areas, the value of r at this point is found from $r^2 = h/n$. Substituting for h/n from (164) gives

(185)
$$r = a(1 - e^2)^{1/4}.$$

The appropriate value of the true anomaly f can then be found from the polar equation of the orbit (169). The value of the eccentric anomaly E

---

[3]The terms *perigee* and *apogee* replace *perihelion* and *aphelion* when the Earth replaces the Sun at the center of force.

[4]There are three other types of libration.

- the axis is not perpendicular to the Moon's orbital plane, causing *libration in latitude;*
- the observer will be moving in space because of the Earth's rotation, causing *diurnal libration;*
- the Moon oscillates slightly because of its nonspherical shape, causing *physical libration.*

can be obtained from the tangent half-angle formula (181). Finally, the mean anomaly M can be calculated by Kepler's equation (188), which we will derive in the next section, and the time of maximum libration in longitude can be found as M/n.

EXERCISE III.2. Compute the maximum libration in longitude and find where in the orbit this is achieved. Prove that if e^2 is neglected in the final answer, then the maximum libration expressed in radians in terms of the period T is

$$(186) \qquad \left(\frac{1}{4} - \frac{5e}{8\pi} \right) T.$$

If the only libration were that in longitude, estimate the percentage of the Moon's surface that could never be seen from the Earth.

EXERCISE III.3. If ψ is the angle between the direction of a planet's motion and the direction perpendicular to the radius vector, show that

$$(187) \qquad \tan \psi = \frac{e \sin E}{\sqrt{1 - e^2}}.$$

EXERCISE III.4. Prove that the greatest value of dr/dt in an elliptic orbit occurs at the ends of the latus rectum, and find this value in terms of a and e.

4. Kepler's equation

Kepler's equation relates the eccentric anomaly E and the mean anomaly M; it is

$$(188) \qquad E - e \sin E = M.$$

We will derive Kepler's equation. From $r = a(1 - e \cos E)$, we find that

$$(189) \qquad \dot{r} = ae \sin E \cdot \dot{E}.$$

From

$$(190) \qquad \frac{1}{r} = \frac{1 + e \cos f}{a(1 - e^2)},$$

we find that

$$(191) \qquad \frac{\dot{r}}{r^2} = \frac{e \sin f \cdot \dot{f}}{a(1 - e^2)}.$$

Because $r^2\dot{f} = h = na^2\sqrt{1 - e^2}$, the last equation becomes

(192)
$$\dot{r} = \frac{nae\sin f}{\sqrt{1 - e^2}}.$$

Comparison of the two representations (189) and (192) of $\dot{r}$ yields

(193)
$$\dot{E} = \frac{n\sin f}{\sqrt{1 - e^2}\,\sin E}.$$

But, from (171), we have

(194)
$$\eta \equiv r\sin f = b\sin E = a\sqrt{1 - e^2}\,\sin E.$$

Hence, $r\dot{E} = na$, or

(195)
$$(1 - \cos E)\dot{E} = n.$$

Integrating, we get

(196)
$$E - e\sin E = nt + c,$$

where c is a constant of integration. But $E = 0$ when $t = \tau$, whence $c = -n\tau$. Then we have

(197)
$$E - e\sin E = M \equiv n(t - \tau),$$

which is Kepler's equation.

EXERCISE III.5. Show that

(198)
$$n(t - \tau) = \sin^{-1}\left\{\frac{\sqrt{1 - e^2}\,\sin f}{1 + e\cos f}\right\} - \frac{e\sqrt{1 - e^2}\,\sin f}{1 + e\cos f}.$$

EXERCISE III.6. We have been working with elliptic motion, where $0 < e < 1$. The case $e = 1$ is parabolic motion.

By working from first principles, or by letting $e \to 1$ in (198), show that parabolic motion must obey the relation

(199)
$$\left(\frac{\mu}{2p^3}\right)^{1/2}(t - \tau) = \tan\tfrac{1}{2}\vartheta + \tfrac{1}{3}\tan^3\tfrac{1}{2}\vartheta.$$

This equation, of the form $x^3 + 3x = c$ and sometimes called Barker's equation, was solved numerically by Edmond Halley, Newton's contemporary and one of the founders of modern numerical analysis.[5] Halley tried to compute a parabolic orbit for the comet of 1682. This entailed solving (199) for about 100 different values of c in the sequence $c = 0.04, 0.08, 0.12, \ldots$. After this attempt failed, Halley was able to fit an elliptical orbit to the observations. He then predicted that the comet would return.[6] There was never any question whether the comet *would* return. That it did so return was a triumph for the 'clockwork' view of the universe and convincing evidence for Newtonian mechanics over various scientific, philosophical, and superstitious theories of the universe competing for favor in Newton's time.

EXERCISE III.7. Prove that the amount of heat received from the Sun by a planet per unit area and in unit time is on the average proportional to the product of the reciprocals of the major and minor axes of its orbit.

5. Solution of Kepler's equation

It would be useful to have all the formulas for the elliptical motion expressed in terms of *one* of the three anomalies f, E, M. A review of the formulas in §3 will make it evident that explicit formulas in terms of f or of E are easily obtained. Kepler's equation (188) is an obstruction to writing the formulas in terms of M. In Chapter IV, we will derive series

---

[5] Although the figure of Edmond Halley (1656–1742) stands forever in the shadow of the Olympian Isaac Newton, and although his popular fame is tied to the periodic return of the heavenly body called Halley's comet, the scope of Halley's scientific achievements is so broad and so deep that he is to be counted among the founders of modern science and the scientific method. Halley established his international reputation through innovative research in the areas of geophysics, navigation, mathematics, and astronomy. He founded physical oceanography, developed the study of geomagnetism, and contributed to meteorology. He devised new ways of charting and proposed a method for discovering the longitude. He made important advances in the practical solution of equations. He discovered stellar motion and was the first to identify a periodic comet. He also made practical contributions to optics and to the design of a diving bell and of various scientific instruments. In the heady days when the Royal Society was young and the notion of 'natural philosopher' had not yet fragmented into a variety of specializations, a talented person could, as Halley did, become influential across the whole spectrum of scientific thought. (See N. Grossman, 'Focus on Halley,' *The Clark Library Bulletin,* Fall, 1985.)

[6] Halley recognized that the returning comet would pass by Jupiter, which would perturb its flight. His confidence in his prediction of the next perihelion weakened over successive published calculations. As late as 1717, he still had to make a 'ballpark' estimate of Jupiter's effect, which led to his hedging prediction that the comet would reach perihelion in late 1758 or early 1759.

Regarding his efforts, Halley wrote that 'all this is nothing but a light trial, and we leave the effort of making this matter deeper to those who survive until the event justifies our predictions.' In a time (such as it yet is) when glory in science was a national glory, Halley boasted that 'Posterity will not forget that *it is to an Englishman that it owes this discovery.*' In 1748, three French scientists closeted themselves away for six months to make a grueling calculation that predicted the perihelion with error of only 31 days. (See the footnote on p. 157.)

expansions for E and for functions of E in terms of M. Unfortunately, there is no elementary closed-form expression for E as a function of M. Here we will consider methods for finding numerical solutions of Kepler's equation to any desired precision.

It seems 'geometrically obvious' that there is exactly one value of E in the interval $[n\pi, (n+1)\pi]$ corresponding to each M in the same interval, when n is an integer. To be rigorous, put

(200) $$F(E) = E - e\sin E - M,$$

where M is fixed in $(n\pi, (n+1)\pi)$—the endpoints offer no problem. Then $F(n\pi) = n\pi - M < 0$ and $F((n+1)\pi) = (n+1)\pi - M > 0$. Because F is continuous, it has a root in the interval $(n\pi, (n+1)\pi)$. Moreover, $F'(E) = 1 - e\cos E > 0$ for all E, because $0 \le e < 1$ for elliptic motion. Therefore, F is monotonic, and there is only one root.

Now we consider an abstraction of this problem. Let φ be a real-valued, differentiable function defined (for convenience) on the whole real line. A *fixed point* of φ is a number x^* such that $x^* = \varphi(x^*)$.

The following procedure, called the *method of iteration*, often locates fixed points. Begin by *guessing* an approximation, x_0, to the fixed point sought. Compute $x_1 = \varphi(x_0)$. If $x_1 = x_0$, then a fixed point has been found. If not, x_1 may be a better approximation to the fixed point. Compute $x_2 = \varphi(x_1)$. Then x_2 may be a better approximation, and so on. Thus we define a sequence $(x_0, x_1, x_2, \dots)$ by the formula

(201) $$x_{n+1} = \varphi(x_n) \quad \text{for } n \ge 0.$$

Suppose that $\lim_{n\to\infty} x_n = \xi$ exists. Because φ is continuous,

(202) $$\begin{aligned} \xi &= \lim_{n\to\infty} x_{n+1} \\ &= \varphi\left(\lim_{n\to\infty} x_n\right) \\ &= \varphi(\xi), \end{aligned}$$

so that ξ must be a fixed point of φ.

No necessary condition is known for the convergence of the iteration sequence to a fixed point. The following is a useful sufficient condition:

If there is a positive constant $k < 1$ such that $|\varphi'(t)| \leq k$ for all t, then the iteration sequence $\{x_n\}$ converges to a unique fixed point ξ, and

$$(203) \qquad |\xi - x_n| < \frac{k^n}{1-k} |\varphi(x_0) - x_0|.$$

The proof of this assertion invokes the Mean Value Theorem in the form

$$(204) \qquad |\varphi(b) - \varphi(a)| \leq \left(\max_{a \leq t \leq b} \varphi'(t) \right) |b - a| \leq k|b - a|.$$

For any $n > 0$,

$$
\begin{aligned}
(205) \qquad |x_n - x_{n-1}| &= |\varphi(x_{n-1}) - \varphi(x_{n-2})| \\
&\leq k|x_{n-1} - x_{n-2}| \\
&\leq k^2|x_{n-2} - x_{n-3}| \\
&\vdots \quad \vdots \\
&\leq k^{n-1}|x_1 - x_0|.
\end{aligned}
$$

Therefore, if $m < n$,

$$
\begin{aligned}
|x_n - x_m| &\leq |x_n - x_{n-1}| + |x_{n-1} - x_{n-2}| + \cdots + |x_{m+1} - x_m| \\
(206) \qquad &\leq [k^{n-1} + \cdots + k^m]|x_1 - x_0|.
\end{aligned}
$$

The sum $k^{n-1} + \cdots + k^m$ is a segment of a geometric series which converges as $n \to \infty$ because $k < 1$. By Cauchy's Criterion, $k^{n-1} + \cdots + k^m \to 0$ as $m \to \infty$ (and so also $n \to \infty$). The iterates therefore form a Cauchy sequence. The real numbers are complete. Hence, the iterates x_n converge to a real number ξ. We have seen that $\varphi(\xi) = \xi$. If η were also a fixed point of φ, then we would have

$$(207) \qquad |\xi - \eta| = |\varphi(\xi) - \varphi(\eta)| \leq |\xi - \eta|.$$

Because $k \leq 1$, we conclude that $|\xi - \eta| = 0$, and so that $\xi = \eta$. Finally, letting $n \to \infty$, we obtain

$$
\begin{aligned}
(208) \qquad |\xi - x_m| &\leq \left[\sum_{r=m}^{\infty} k^r \right] |x_1 - x_0| \\
&\leq \frac{k^m}{1-k} |x_1 - x_0|.
\end{aligned}
$$

It often occurs in practice that the hypotheses are not satisfied for all t. It is enough that there be an interval $[a, b]$ such that $\varphi(t)$ is in $[a, b]$ whenever t is in $[a, b]$ and that $|\varphi'(t)| \leq k < 1$ for all t in $[a, b]$. There may be several such intervals disjoint from each other and so several distinct fixed points.

Application to Kepler's equation (188) is immediate. Let $\varphi(E) = M + e \sin E$ for every real E. Then $\varphi'(t) = e \cos E$, so that $|\varphi'(t)| \leq e < 1$ for all E. The sufficient condition for convergence is satisfied, and we may solve Kepler's equation by iteration. There are tables of E for various values of e and M, and these can be used to get starting guesses E_0. However, even nonprogrammable, scientific, 'pocket' calculators give a handy way to do the calculations. In astronomical problems, e is usually relatively small, seldom bigger than a few tenths. Hence, convergence is rapid. You must remember to work in radians or else to convert e into degrees as $e° = 180e/\pi$, in which case Kepler's equation becomes

$$(209) \qquad\qquad E° - e° \sin E° = M°.$$

Of course, *automatic* root finders are now available, even on some 'low-end' pocket calculators.

EXERCISE III.8. Solve Kepler's equation by iteration when $M = 45°$ and $e = 0.2$. Carry out enough iterations to get some indication of the rate of convergence, and estimate your error.

EXERCISE III.9. Use iteration in Kepler's equation in the form

$$(210) \qquad\qquad E_{i+1} = E_i + e \sin E_i, \quad E_0 = M.$$

Neglecting e^3 and higher powers, show that

$$(211) \qquad\qquad E_2 = M + e \sin M + e^2 \sin M \cos M.$$

EXERCISE III.10. Show that if the eccentricity e is so small that e^3 may be neglected, then

$$(212) \qquad\qquad r = a(1 - e \cos nt + e^2 \sin^2 nt)$$

and

$$(213) \qquad\qquad \omega = n(1 + \tfrac{1}{2}e^2 \cos 2nt),$$

where r is the radius, t the time measured from pericenter ($\tau = 0$), $2\pi/n$ the period, a the semi-major axis, and ω the angular velocity around the *empty* focus.

Hence prove that the Moon very nearly turns the same face to that focus of its orbit in which the Earth is not situated.

6. The velocity of a planet in its orbit

Let v be the velocity of a planet in its orbit. Then, by (171),

$$(214) \qquad v^2 = \dot{\xi}^2 + \dot{\eta}^2.$$

We know from (175) that $\xi = a(\cos E - e)$ and $\eta = a\sqrt{1-e^2}\,\sin E$; hence

$$(215) \qquad v^2 = a^2(1 - e^2 \cos^2 E)\dot{E}^2.$$

Using Kepler's equation, we get

$$(216) \qquad (1 - e \cos E)\dot{E} = n.$$

Therefore

$$(217) \qquad v^2 = n^2 a^2 \frac{1 + e \cos E}{1 - e \cos E} = \frac{\mu}{r}[2 - (1 - e \cos E)],$$

or

$$(218) \qquad v^2 = \mu\left(\frac{2}{r} - \frac{1}{a}\right),$$

the formula required. This equation clearly expresses conservation of energy.

EXERCISE III.11. A particle acted upon by a central attraction μ/r^2 is projected with a velocity u at right angles to the radius vector at a distance c from the center of force. Investigate the limits of u that yield an ellipse as orbit, and in case of elliptic orbit find the eccentricity and the major axis, and also find the condition as to whether the projection point is at the end of the major axis at pericentron or apocentron.

EXERCISE III.12. Show that in elliptic motion under Newton's Law of Universal Gravitation, the quantity $\int K\,dt$, where K denotes kinetic energy, integrated over a complete period, depends upon only the semi-major axis and not on the eccentricity.

EXERCISE III.13. When a periodic comet is at its greatest distance from the Sun, its speed v receives a small increment δv. Show that the comet's least distance from the Sun will be increased by the quantity

$$(219) \qquad 4\,\delta v\sqrt{\frac{a^3}{\mu} \cdot \frac{1-e}{1+e}}.$$

EXERCISE III.14. If PP_0P' is a focal chord of an elliptical path around the Sun, then show that the time from P to P' through perihelion is equal to the time of falling toward the Sun from a distance $2a$ to a distance $a(1 + \cos \alpha)$, where $\alpha = 2\pi - (E' - E)$, and $E' - E$ is the difference of the eccentric anomalies of the points P, P'.

EXERCISE III.15. At a certain point in an elliptical orbit described under a central force μ/r^2, the constant μ is suddenly changed by a small amount. If the eccentricities of the new and the old orbits are the same, show that the point must be at the end of the minor axis.

EXERCISE III.16. A planet describing a circular orbit receives a small impulse in the direction of its motion. At any future time it is at P, whereas if it had not received the impulse it would have been at P'. Show that although the path of P is never far from the original circle, the length of PP' need not be small.

Discuss the case where the small impulse is radial.

EXERCISE III.17. A particle is describing an ellipse of eccentricity e under the action of an inverse-square attractor toward a focus. When it is at the pericenter, the center of force is transferred to the other focus. Prove that the eccentricity of the new elliptical orbit is

(220)
$$\frac{e(1 + e)}{1 - e}.$$

7. Drifting of the gravitational constant

In 1937, P.A.M. Dirac formulated the so-called *large-numbers hypothesis*.[7] He observed that when the fundamental units of mass, length, and time are taken to be the mass of the electron, the radius of the electron's orbit in the hydrogen atom, and the time taken for a photon to traverse that radius, respectively, many physical constants have values that are about 1 within a few orders of magnitude. For example, the mass of the proton is 1836 (about 10^3), the speed of light is 1, and the fine-structure constant of the hydrogen atom is about $1/137$ (about 10^{-2}). There are various other examples. On the other hand, the age of the universe is about 10^{40} while the gravitational constant G is about 10^{-40}. Can it be a coincidence that the value of G in the chosen units is the reciprocal of the age of the

---

[7]For general background and further references, see Thomas C. van Flandern, 'Is Gravity Getting Weaker?', *Scientific American*, 1976.

universe? Or does it represent some as yet undiscovered relation between the gravitational constant and the age of the universe?

This smacks of the 'Principle of Correspondences' and other magical syllogisms that lie at the root of belief in astrology,[8] but Dirac took the coincidence seriously. He pointed out that one of two alternatives must follow. The first is that because the age of the universe increases with time, the value of G must also decrease with time. The second is that the fundamental units themselves could be changing in time. If we want to investigate the consequences of the hypothesis that G decreases with time, we must realize that this has only recently become possible. Up to 1955, the only way to measure time was with a pendulum clock or by ephemeris time, the time scale determined by the revolution of the Earth around the Sun. Both of these scales involve G directly, so they can not be used to compare changes of G to changes of period. Since 1955, atomic clocks of sufficient precision have become available. If the fundamental units do not change, then atomic clocks should keep a rate constant with respect to changes in G. Therefore, it should be possible to investigate changes in the periods of celestial bodies to look for changes in G.

Such investigations have been carried out. As explained in van Flandern's article, observations of occultations of stars by the moon can yield very precise values of the changes (if any) in G. Furthermore, the Moon will slowly recede from the Earth. Very precise measurements of the distance to the Moon were made before 1976 by bouncing laser beams off corner reflectors placed on the Moon. As analyzed by astronomers at the University of Texas, with all known gravitational effects and perturbations subtracted off, the observations seemed to indicate that G is decreasing. Of course, there may yet be an undiscovered 'ordinary' cause. It is interesting to note that Einstein's General Theory of Relativity predicts that G is constant.

We are going to work out the change in period and radius due to a change in G, assuming that masses, times, lengths, etc. are measured in

---

[8]B.J. Bok and L.E. Jerome, *Objections to Astrology*, Prometheus Press, Buffalo, N.Y., 1975.

unchanging units. Suppose that

(221) $$G = G_0(1 - \epsilon t),$$

where ϵ is so small that ϵ^2 can be neglected over the time period considered. If we do not insist upon times too large, then we can as well write

(222) $$G = G_0 e^{-\epsilon t},$$

whence

(223) $$\frac{\dot{G}}{G} = -\epsilon.$$

Therefore, ϵ is the relative decrease in G with time. If M is the mass of the attracting center, then the attracting force has magnitude GM/r^2. Setting $\mu = GM$, we have

(224) $$\frac{\dot{\mu}}{\mu} = \frac{\dot{G}}{G} = -\epsilon.$$

To avoid the excesses of subscripts, we will write the magnitude of the force as

(225) $$\frac{\mu}{r^2}(1 - \epsilon t),$$

with μ constant.

There is a substantial simplification of the calculations with no loss of physical meaning that can be gained by supposing the Moon to be going around the Earth in a circular orbit of radius $r_0 = h^2/\mu$, where h is the angular momentum constant.

Because the force is always central, the angular momentum remains constant in time. Therefore, the radial equation of motion (89) becomes

(226) $$\ddot{r} - \frac{h^2}{r^3} = -\frac{\mu}{r^2}(1 - \epsilon t).$$

It is not possible to use the equation of areas to transform the independent variable to an angle because t is explicit in this equation. The transformation of the dependent variable by $r = 1/u$ is of no use. We use a perturbation expansion.

If $\epsilon \to 0$, the differential equation describes the circular motion. When ϵ is 'small,' the change in the differential equation is 'small.' The solutions of 'nice' differential equations depend continuously upon the coefficients, so we expect the change in r to be 'small.' Set

(227) $$r = \frac{h^2}{\mu}(1 + \epsilon y(t)),$$

where the function y is to be determined. Neglecting ϵ^2,

(228) $$r^2 = \frac{h^4}{\mu^2}(1 + 2\epsilon y), \qquad \frac{1}{r^2} = \frac{\mu^2}{h^4}(1 - 2\epsilon y),$$

$$r^3 = \frac{h^6}{\mu^3}(1 + 3\epsilon y), \qquad \frac{1}{r^3} = \frac{\mu^3}{h^6}(1 - 3\epsilon y).$$

Plugging into the differential equation leads to

(229) $$\frac{h^2}{\mu}\ddot{y}\epsilon - \frac{h^2\mu^3}{h^6}(1 - 3\epsilon y) = -\frac{\mu^3}{h^4}(1 - 2\epsilon y)(1 - \epsilon t).$$

The two sides of this differential equation can be considered to be (quadratic) polynomials in ϵ which must agree up to and including the terms in ϵ^1, because ϵ^2 has been neglected. The terms in ϵ^0 correspond to putting $\epsilon = 0$, resulting in a trivial identity. Equating the coefficients of ϵ gives the differential equation

(230) $$\frac{h^2}{\mu}\ddot{y} + \frac{3\mu^3}{h^4}y = \frac{2\mu^3}{h^4}y + \frac{\mu^3}{h^4}t.$$

After some simple algebra, we get a differential equation to solve for y:

(231) $$\ddot{y} + \frac{\mu^4}{h^6}y = \frac{\mu^4}{h^6}t.$$

This differential equation is easy to solve; the general solution is

(232) $$y = t + A\sin\frac{\mu^2}{h^3}t + B\cos\frac{\mu^2}{h^3}t,$$

where A and B are constants. Put $T_0 = 2\pi h^3/\mu^2$, the period of the circular motion. Then

(233) $$y = t + A\sin\frac{2\pi t}{T_0} + B\cos\frac{2\pi t}{T_0}.$$

If G is decreasing, then the decrease probably didn't start in recent times. Hence, we may suppose that the effect of the periodic terms has long since ceased to be significant compared to the increasing term t. In any event, the average of the periodic terms over a period T_0 is 0. We are justified in using

$$(234) \qquad\qquad r = \frac{h^2}{\mu}(1 + \epsilon t).$$

Therefore, the relative change in G can be determined from observations of r.

Let T_ϵ be the period of the Moon when it is viewed against the fixed stars. The changes in radius, period, and gravitational constant are very slow. Therefore, we start with Kepler's Third Law,

$$(235) \qquad\qquad \frac{a^3}{T^2} = \frac{\mu}{4\pi^2}.$$

Take variations of this equation with respect to ϵ, using δ to indicate the variations. Then

$$(236) \qquad\qquad 3\frac{\delta a}{a} - 2\frac{\delta T}{T} = \frac{\delta \mu}{\mu}.$$

Identify r with a, rewriting the equation (234) as the approximation

$$(237) \qquad\qquad r = \frac{h^2}{\mu}e^{\epsilon t},$$

valid if ϵ^2 is neglected. Then $\delta T/T = \epsilon\,\delta t$. Similarly, $\delta\mu/\mu = -\epsilon\,\delta t$, whence $\delta T/T = 2\epsilon\,\delta t$. Therefore,

$$(238) \qquad\qquad T_\epsilon = T_0(1 + 2\epsilon t),$$

so that the increase in the period with time is proportional to $2\epsilon = -2\dot{G}/G$.

Observations of the occultations of fixed stars by the Moon, when corrected for all known 'ordinary' causes, lead to the conclusion that G is decreasing by 3.6 ± 1.8 parts in 10^{11} per year.[9]

EXERCISE III.18. If the change in G is as just quoted, by how many seconds is the period of the Moon increasing during each revolution?

EXERCISE III.19. A simple pendulum beats at the surface of the Earth with infinitesimal amplitude. What is the effect produced on its period by the supposed drift in G?

---

[9]This is under the 'primitive' theory. Under the second alternative, that the fundamental units are changing, the decrease in g is twice as large.

Expansions for an Elliptic Orbit

1. The general problem

We have studied motion in an elliptic orbit under the inverse square law of attraction to a center. Among the formulas we derived in §III.3 and (196) were these:

$$(247) \qquad r = \frac{a(1 - e^2)}{1 + e \cos f},$$

$$(248) \qquad \tan \tfrac{1}{2} f = \sqrt{\frac{1 + e}{1 - e}} \, \tan \tfrac{1}{2} E,$$

$$(249) \qquad r = a(1 - e \cos E),$$

$$(250) \qquad E - e \sin E = M.$$

In planetary and satellite theory, e is generally small. Therefore, it is possible to express functions of one or several of the variables r, f, E, M in terms of one of the remaining variables. For applications, it is often desired to express various functions in terms of the time; that is, in terms of $M = n(t - \tau)$. Because f and E are periodic functions of M, the expansions are often of a form that exploits periodicity. These are Fourier series. Other types of expansions arise. We will first describe the various expansion techniques in the next few sections. Then we will derive expansions specific to elliptic motion.

2. Lagrange's expansion theorem

Suppose that x and y are variables related by the equation

$$(243) \qquad\qquad y = x + \alpha\varphi(y),$$

where φ is a 'nice' function and α is a 'small' parameter. The problem is to express y (and $F(y)$, where F is a 'nice' function) in terms of x and α.

EXAMPLE. Consider the equation

$$(244) \qquad\qquad y = x + \alpha y^2.$$

Of course, this is a quadratic equation for y, and the two solutions are available by way of the quadratic formula. Having a formula available for the solution is hardly typical of the equations that present themselves to us, so we will ignore the quadratic formula for now and proceed in a typical fashion toward the solution(s).

The form of (244) makes it easy for us to estimate the two solutions for a given pair (x, α) when the 'ordinary' number $x \neq 0$ and α is 'small.' If y is an 'ordinary' number, then we neglect αy^2 to find a value for $y \approx x$. If y is to be quite large, then we neglect x and conclude that, essentially, $y = \alpha y^2$. This truncated equation has the two roots 0 and $1/\alpha$, but we discard the root 0 because it is not 'large.' Now we must refine our first estimates.

Take first the case where $y \approx x$. The equation (244) is in a form that makes it easy to look for a refined estimate of the solution by the iteration process explained in §5. We suppose that $y_{n+1} = x + \alpha y_n^2$ and begin with the guess $y_0 = x$. We use simple algebra to tabulate our successive approximations:

$$\begin{aligned}
y_0 &= x, \\
y_1 &= x + \alpha x^2, \\
y_2 &= x + \alpha x^2 + 2\alpha^2 x^3 + \cdots, \\
&\vdots \quad \vdots \quad \vdots
\end{aligned}$$

The y_2-formula has been truncated by lopping off the α^3-term. Ordinarily, an iteration scheme such as we are carrying out will add one more 'correct' power of the small parameter with each iteration. Extra, higher powers may be carried, but their coefficients will be revised in later steps. In general, the additional algebra that must be carried does not pay off.

What about convergence? The sufficient condition on p. 54 can be applied when suitable changes are made to bring the statement from the number realm into the function realm, and we may conclude that the iterates y_n will converge in some reasonable sense to a solution function $y(x, \alpha)$ in the form of a power series in α with coefficients that are functions of x.

EXERCISE IV.1. Construct an iteration scheme that will produce the root of (244) near $1/\alpha$, and work out the first two approximations from the first guess $1/\alpha$. Look for expansions in powers of $1/\alpha$, and do not require the iteration scheme to look like that of the first case.

Consider the function

(245)
$$G(x, y) = y - x - \alpha\varphi(y).$$

According to the Implicit Function Theorem,[1] the condition $\partial G/\partial y \neq 0$ at a point (x_0, y_0) is sufficient for the equation $G(x, y) = 0$ to have a solution of the form $y = f(x, \alpha)$ in some neighborhood of (x_0, y_0). In the present case, $\partial G/\partial y = 1 - \alpha\varphi'(y)$. If φ' is a bounded function—which is certainly the case when φ' is periodic—then $\partial G/\partial y \neq 0$ for all y, provided only that α is small enough: if $|\varphi'(y)| \leq L$, then it is sufficient that $|\alpha| < 1/L$.

We are led to conjecture that there may be an expansion of a solution to (243) as a power series in α,

(246)
$$y = f(x, \alpha) = x + \alpha\varphi(x) + \sum_{n=2}^{\infty} \Phi_n(x)\alpha^n,$$

where $\Phi_n(x)$ is a universal function of φ and its derivatives of order up to and including $n - 1$. Such an expansion is known.

Lagrange's Expansion Theorem: Let φ and F have derivatives of all orders. Suppose x, y, and α to satisfy the relation

(247)
$$y = x + \alpha\varphi(y).$$

---

[1] T.M. Apostol, *Mathematical Analysis, Second Edition,* Theorem 13.7.

Then there is an expansion

(248) $F(y) = F(x) + \dfrac{\alpha}{1!}[\varphi(x)F'(x)]$

$\qquad\qquad + \dfrac{\alpha^2}{2!}\dfrac{d}{dx}[\varphi(x)^2 F'(x)] + \cdots$

$\qquad\qquad + \dfrac{\alpha^n}{n!}\dfrac{d^{n-1}}{dx^{n-1}}[\varphi(x)^n F'(x)] + \cdots ,$

valid for small enough values of α.

The true convergence picture becomes clear only when complex variables and analytic functions are used.[2]

Not wanting to assume knowledge of analytic function theory, we present a 'symbolic' proof of Lagrange's expansion theorem.[3]

We have supposed that $y = f(x, \alpha)$. Then $F(y)$ can be expressed as a function of x and α, namely as $F(f(x, \alpha))$. Fix x and consider $F(y)$ as a function of α. As such, there is a Maclaurin expansion

(249)

$$F(y) = F|_0 + \frac{\alpha}{1!}\frac{\partial F}{\partial \alpha}\bigg|_0 + \frac{\alpha^2}{2!}\frac{\partial^2 F}{\partial \alpha^2}\bigg|_0 + \cdots + \frac{\alpha^n}{n!}\frac{\partial^n F}{\partial \alpha^n}\bigg|_0 + \cdots ,$$

where the subscript indicates that α is to be set to 0 after the differentiation.

Let A denote the operator $\partial/\partial\alpha$. The preceding expansion can be written as

(250)

$$F(y) = F|_0 + \frac{\alpha}{1!}(AF)|_0 + \frac{\alpha^2}{2!}(A^2 F)\big|_0 + \cdots + \frac{\alpha^n}{n!}(A^n F)|_0 + \cdots .$$

Remembering that $Ax = 0$, we apply A to the equation $y = x + \alpha\varphi(y)$ to get

(251) $Ay = \varphi(y) + \alpha\varphi'(y)Ay.$

Let D denote the operator $\partial/\partial x$. Then

(252) $Dy = 1 + \alpha\varphi'(y)Dy.$

---

[2] E.T. Whittaker and G.N. Watson, *Modern Analysis, Fourth Edition*, §7.32 and p. 149.
[3] W.S. Smart, *Celestial Mechanics*, §3.02.

Multiply the last equation by $\varphi(y)$ and subtract the result from its predecessor. After simple factoring, we find that

$$(253) \qquad (Ay - \varphi\, Dy)(1 - \alpha\varphi'(y)) = 0.$$

We explained above why it is reasonable to assume that $1 - \alpha\varphi'(y) \neq 0$. In consequence,

$$(254) \qquad Ay = \varphi Dy.$$

We conclude that $AF(y) = F'(y)A(y) = \varphi(y)F'(y)Dy$ and $DF(y) = F'(y)Dy$. Hence

$$(255) \qquad AF = \varphi DF.$$

Now we use induction to prove the formula

$$(256) \qquad A^n F = D^{n-1}(\varphi^n DF).$$

The case $n = 1$ has just been verified. Suppose that the nth case is true. The variables x and α are independent. Therefore, the partial derivatives $\partial/\partial x$ and $\partial/\partial\alpha$ commute: $DA = AD$. It follows that

$$(257) \quad A^{n+1}F = AA^n F = AD^{n-1}(\varphi^n DF) = D^{n-1}[A(\varphi^n DF)].$$

When applied to functions of y, $A = \varphi D$. Therefore,

$$(258) \qquad
\begin{aligned}
A(\varphi^n DF) &= \varphi^n D(AF) + A(\varphi^n) \cdot DF \\
&= \varphi^n D(AF) + \varphi D(\varphi^n) \cdot DF \\
&= \varphi^n D(\varphi DF) + \varphi DF \cdot D(\varphi^n) \\
&= D(\varphi^{n+1} DF).
\end{aligned}$$

Hence,

$$(259) \qquad A^{n+1}F = D^n(\varphi^{n+1}DF),$$

which was to be shown. Accordingly,

$$(260) \qquad A^n F = D^{n-1}(\varphi^n DF)$$

for every integer $n \geq 1$.

Translating back into derivative notion, we find we have proved that

$$\text{(261)} \qquad \left.\frac{\partial^n F}{\partial \alpha^n}\right|_0 = \frac{d^{n-1}}{dx^{n-1}}[\varphi(x)^n F'(x)].$$

Inserting this expression into the Maclaurin expansion (249), we arrive at Lagrange's Expansion Theorem.

EXAMPLE. The equation $y = x + \alpha/y$ is a quadratic equation in y with $\varphi(y) = 1/y$. Take $F(y) = y$. We must compute

$$\text{(262)} \qquad \frac{d^{n-1}}{dx^{n-1}}[\varphi(x)^n F'(x)] = \frac{d^{n-1}}{dx^{n-1}}[x^{-n}] = \frac{(-1)^{n-1}(2n-2)!}{(n-1)!\, x^{2n-1}}.$$

Therefore,

$$\text{(263)} \qquad y = x + \sum_{n=1}^{\infty} \frac{(-1)^{n-1}(2n-2)!\, \alpha^n}{(n-1)!\, x^{2n-1}}\frac{}{n!}.$$

Of course, the quadratic equation has the *two* roots

$$\text{(264)} \qquad \frac{x}{2}\left\{1 + \sqrt{1 + \frac{4\alpha}{x^2}}\right\} \quad \text{and} \quad \frac{x}{2}\left\{1 + \sqrt{1 - \frac{4\alpha}{x^2}}\right\}.$$

The expansion converges to the first of these (and the reason is contained in the precise statement of the conditions for convergence).

EXERCISE IV.2. Let z be that root of the equation $z^3 - z^2 - w = 0$ that $\to 1$ as $w \to 0$. Show that

$$\text{(265)} \qquad z = 1 + \sum_{n=1}^{\infty} \frac{(-1)^{n-1}(3n-2)!}{n!(2n-1)!}w^n,$$

and that the series converges for all w with $|w| < 4/27$. [The value $4/27$ is that for which the cubic has a double root. The convergence to z for all w with $|w| < 4/27$ follows from the Inverse Function Theorem for analytic functions.]

EXERCISE IV.3. The equation $ze^{-z} = w$ has a solution

$$\text{(266)} \qquad z = h(w) = w + \frac{2w}{2!} + \cdots + \frac{n^{n-1}}{n!}w^n + \cdots;$$

furthermore,

$$\text{(267)} \qquad e^{\alpha h(w)} = 1 + \sum_{n=1}^{\infty} \frac{\alpha(\alpha + n)^{n-1}}{n!}w^n.$$

Both series converge for $|w| < 1/e$.

3. Bessel coefficients

Consider the generating function

$$U(x, z) = e^{\frac{1}{2}x(z-1/z)}; \tag{268}$$

it can be expanded into a Laurent series

$$U(x, z) = \sum_{n=-\infty}^{\infty} J_n(x)z^n, \tag{269}$$

the coefficients $J_n(x)$ being defined by this expansion. They are called the *Bessel coefficients* of order n.[4]

To display explicit formulas for the Bessel coefficients, we write

$$U(x, z) = e^{\frac{1}{2}xz}e^{-\frac{1}{2}xz^{-1}} \tag{270}$$

$$= \sum_{p=0}^{\infty} \frac{(x/2)^p z^p}{p!} \cdot \sum_{q=0}^{\infty} \frac{(-x/2)^q z^{-q}}{q!}. \tag{271}$$

Multiply the two series together and rearrange the individual products by powers of z, justifying the rearrangement by an appeal to absolute convergence. First take $n \geq 0$. Then we want to match the power z^{-r} with the power z^{r+n} to get the product z^n. Gathering the terms, we get

$$J_n(x) = \sum_{r=0}^{\infty} \frac{(-1)^r (x/2)^{2r+n}}{r!\,(r+n)!}. \tag{272}$$

This formula does not work for $n < 0$, for then it requires division by 0. However, we note that

$$U(x, -1/z) = U(x, z). \tag{273}$$

By definition,

$$U(x, -1/z) = \sum_{n=-\infty}^{\infty} J_n(x)(-1/z)^n \tag{274}$$

$$= \sum_{n=-\infty}^{\infty} (-1)^n J_{-n}(x)z^n. \tag{275}$$

---

[4]Sometimes they are called the *Bessel functions* of order n, because it is possible to define $J_n(x)$ for every complex number n by means of the differential equation (287) to be obtained below.

Comparing this result with (269) and invoking the uniqueness of Laurent representations, we get

(276) $$J_{-n}(x) = (-1)^n J_n(x).$$

From the defining formula (268),

(277) $$\frac{\partial U}{\partial z} = \frac{x}{2}(1 + z^{-2})U,$$

and, from the expansion (269),

(278) $$\frac{\partial U}{\partial z} = \sum_{n=-\infty}^{\infty} n J_n(x) z^{n-1} = \sum_{n=-\infty}^{\infty} (n+1) J_{n+1}(x) z^n.$$

Hence,

(279) $$\frac{x}{2}(1 + z^{-2}) \sum_{n=-\infty}^{\infty} J_n(x) z^n = \sum_{n=-\infty}^{\infty} (n+1) J_{n+1}(x) z^n.$$

Equating coefficients of z^n, we find that

(280) $$\tfrac{1}{2}x\{J_n(x) + J_{n+2}(x)\} = (n+1) J_{n+1}(x),$$

or, writing $n - 1$ for n, we get the symmetrical

(281) $$n J_n(x) = \frac{x}{2}\{J_{n-1}(x) + J_{n+1}(x)\}.$$

We have

(282) $$\frac{\partial U}{\partial x} = \tfrac{1}{2}(z - z^{-1})U = \sum_{n=-\infty}^{\infty} J_n'(x) z^n,$$

or

(283) $$\tfrac{1}{2}(z - z^{-1}) \sum_{n=-\infty}^{\infty} J_n(x) z^n = \sum_{n=-\infty}^{\infty} J_n'(x) z^n.$$

Equating the coefficients of z^n gives

(284) $$J_n'(x) = \tfrac{1}{2}\{J_{n-1}(x) - J_{n+1}(x)\}.$$

Now we will show that $J_n(x)$ satisfies a certain second-order differential equation. Differentiate the last equation and plow it back into the result:

$$
\begin{aligned}
J_n''(x) &= \tfrac{1}{2}\{J_{n-1}'(x) - J_{n+1}'(x)\} \\
(285) \qquad &= \tfrac{1}{4}\{[J_{n-2}(x) - J_n(x)] - [J_n(x) - J_{n+2}(x)]\} \\
&= -J_n(x) + \tfrac{1}{4}[J_{n-2}(x) + J_n(x)] + \tfrac{1}{4}[J_n(x) + J_{n+2}(x)]\}.
\end{aligned}
$$

Using (281), we find that

$$(286)\quad J_n''(x) + J_n(x) = \frac{1}{2x}\{(n-1)J_{n-1}(x) + (n+1)J_{n+1}(x)\}$$

$$= \frac{n^2}{x^2} J_n(x) - \frac{1}{x} J_n'(x).$$

By this calculation, we have verified that $J_n(x)$ satisfies *Bessel's differential equation of order n*:

$$(287)\qquad \frac{d^2y}{dx^2} + \frac{1}{x}\frac{dy}{dx} + \left(1 - \frac{n^2}{x^2}\right)y = 0.$$

Finally, we will derive an integral representation for the Bessel coefficient $J_n(x)$. Put $z = e^{i\vartheta}$. Then $\frac{1}{2}(z - z^{-1}) = i\sin\vartheta$. Therefore

$$(288)\qquad U \equiv e^{ix\sin\vartheta} = \sum_{p=-\infty}^{\infty} J_p(x)e^{ip\vartheta}.$$

An appeal to uniform convergence will justify multiplication of this series by $e^{-in\vartheta}$ and term-by-term integration from $\vartheta = 0$ to $\vartheta = 2\pi$. Because

$$(289)\qquad \int_0^{2\pi} e^{ik\vartheta}\,d\vartheta = \begin{cases} 2\pi, & \text{if } k = 0, \\ 0, & \text{if } k \neq 0, \end{cases}$$

we obtain

$$(290)\qquad 2\pi J_n(x) = \int_0^{2\pi} e^{-i(n\vartheta - x\sin\vartheta)}\,d\vartheta.$$

Write the integral as the sum of two, one with integration from 0 to π and the second from π to 2π. In the second, replace ϑ by $2\pi - \vartheta$. Then

$$(291)\quad 2\pi J_n(x) = \int_0^{\pi} e^{-i(n\vartheta - x\sin\vartheta)}\,d\vartheta + e^{-2in\pi}\int_0^{\pi} e^{i(n\vartheta - x\sin\vartheta)}\,d\vartheta.$$

But $e^{-2in\pi} = 1$ and

$$(292)\qquad e^{i(n\vartheta - x\sin\vartheta)} + e^{-i(n\vartheta - x\sin\vartheta)} = 2\cos(n\vartheta - x\sin\vartheta).$$

We arrive, therefore, at *Bessel's integral*:

$$(293)\qquad J_n(x) = \frac{1}{\pi}\int_0^{\pi} \cos(n\vartheta - x\sin\vartheta)\,d\vartheta.$$

It can be verified directly by differentiation under the integral sign that the integral in (293) satisfies the Bessel differential equation (287) *if n is an integer.* The integral must be modified to represent $J_p(x)$ if p is not an integer.

EXERCISE IV.4. Verify these properties of the Bessel coefficients.

(a) $\{x^n J_n(x)\}' = x^n J_{n-1}(x)$.

(b) $J_0'(x) = -J_1(x)$.

(c) $\{x^{-n} J_n(x)\}' = -x^{-n} J_{n+1}(x)$.

(d) $|J_n(x)| \leq 1$ for all real x.

(e) By multiplying the expansions for $\exp[(x/2)(z - z^{-1})]$ and $\exp[-(x/2)(z - z^{-1})]$ and considering the terms independent of z, show that

$$J_0(x)^2 + 2J_1(x)^2 + 2J_2(x)^2 + \cdots = 1.$$

Deduce that $|J_0(x)| \leq 1$ and $|J_n(x)| \leq 1/\sqrt{2}$ if $n \geq 1$, both for all x.

4. Fourier series

Suppose a real-valued function f were defined for real numbers and had period 2π: $f(x + 2\pi) = f(x)$ for all x. It occurred to Euler and to others to investigate whether f could be synthesized from the most elementary smooth functions of period 2π. Thus, Euler sought an expansion

$$(294) \qquad f(x) = \tfrac{1}{2}a_0 + \sum_{n=1}^{\infty} (a_n \cos nx + b_n \sin nx),$$

where the constants a_n and b_n are to be determined in terms of the properties of f. Such expansions are called *Fourier series* to honor J. Fourier, whose book *Théorie de la Chaleur*,[5] published in 1822, was a milestone in the application of mathematics to physics. Curiously enough, Fourier used a laborious method for determining the coefficients, one that required solving infinitely many linear equations in infinitely many unknowns, even though he mentions the generally simpler method of Euler that we now will describe.

Suppose that it is permissible to multiply the series (294) by $\cos kx$ and to integrate the resulting series term-by-term for $0 \leq x \leq 2\pi$. It is easy to show that

$$(295) \qquad \int_0^{2\pi} \begin{Bmatrix} \cos \\ \sin \end{Bmatrix} kx \begin{Bmatrix} \cos \\ \sin \end{Bmatrix} nx \, dx = 0 \quad \text{for } k \neq n,$$

---

[5] Available in English with the title *The Analytic Theory Of Heat*, Dover, 1955, and worth looking into.

and even for $k = n$ if the functions disagree. Furthermore,

$$(296) \qquad \int_0^{2\pi} \cos^2 nx \, dx = \int_0^{2\pi} \sin^2 nx \, dx = \pi.$$

Therefore, the coefficients will be determined by the *Euler formulas:*

$$(297) \qquad a_n = \frac{1}{\pi} \int_0^{2\pi} f(x) \cos nx \, dx, \quad (n = 0, 1, 2, 3, \dots),$$

$$(298) \qquad b_n = \frac{1}{\pi} \int_0^{2\pi} f(x) \sin nx \, dx, \quad (n = 1, 2, 3, \dots).$$

As a simple consequence of periodicity, the integration can be carried out over any interval of length 2π.

Two questions arise immediately. First, if f does indeed possess a Fourier expansion (294), does this method lead to the (or a) correct expansion? (That is, is term-by-term integration permissible?) Second, if the coefficients are determined by the Euler formulas (297) and the series (294) is formed, does it converge to f or, indeed, converge at all? The investigations required to answer these questions, insofar as they have been answered, are among the most subtle and profound in all of mathematics. At the same time, they are among the most beautiful. A major part of the modern theory of integration is a by-product of the investigation of Fourier series. The integrals of Riemann and then of Lebesgue, of Perron, and of Denjoy, among others, were formulated under this impetus.[6] Fortunately, those functions with which we will deal are so nicely behaved that the earliest and simplest of the general sufficient conditions for the validity of the Fourier series expansion will do for us.

> *Dirichlet's Condition.* Suppose that f is bounded and has only a finite number of maxima, minima, and discontinuities in any bounded interval. Then the Fourier series
>
> $$(299) \quad f(x) = \tfrac{1}{2}a_0 + \sum_{n=1}^{\infty} (a_n \cos nx + b_n \sin nx),$$

---

[6]T. Hawkins, *Lebesgue's Theory Of Integration.* This book is a very readable history of the successes *and* the failures of a substantial part of nineteenth-century mathematics.

where

(300)
$$a_n = \frac{1}{\pi} \int_0^{2\pi} f(x) \cos nx \, dx, \quad b_n = \frac{1}{\pi} \int_0^{2\pi} f(x) \sin nx \, dx,$$

converges to

(301)
$$\frac{f(x-0) + f(x+0)}{2}.$$

In every example that we compute, the validity of Dirichlet's Condition will be obvious or nearly so.

It is often possible—and convenient—to obtain a Fourier expansion by *ad hoc* methods rather than from the integral formulas (297) for the coefficients. This is permissible because the Fourier series of a function is unique if indeed it exists.

Manipulation of complex numbers offers a convenient technique for working out Fourier series. We follow ordinary mathematical usage by letting $i = \sqrt{-1}$. Euler discovered the relation

(302)
$$e^{ix} = \cos x + i \sin x,$$

from which follows

(303)
$$e^{-ix} = \cos x - i \sin x$$

and, by algebra,

(304)
$$\cos x = \frac{e^{ix} + e^{-ix}}{2} \quad \text{and} \quad \sin x = \frac{e^{ix} - e^{-ix}}{2i}.$$

Also,

(305)
$$e^{inx} = \cos nx + i \sin nx,$$

with three further equations as above.

Formal manipulation gives

$$(306) \quad \frac{1}{2}a_0 + \sum_{n=1}^{\infty} [a_n \cos nx + b_n \sin nx]$$

$$= \frac{1}{2}a_0 + \sum_{n=1}^{\infty} \left[a_n \left(\frac{e^{inx} + e^{-inx}}{2} \right) + b_n \left(\frac{e^{inx} - e^{-inx}}{2i} \right) \right]$$

$$= \sum_{n=-\infty}^{\infty} c_n e^{inx},$$

where

$$(307) \quad c_n = \frac{1}{2}(a_n - ib_n) = \frac{1}{2\pi} \int_0^{2\pi} f(x) e^{-inx}\, dx$$

and $c_{-n} = \overline{c_n}$ (complex conjugate).

5. Preliminaries for expansions

Introduce variables ξ and η related to the true anomaly f and the eccentric anomaly E by

$$(308) \quad \xi = e^{if} \quad \text{and} \quad \eta = e^{iE},$$

where $i = \sqrt{-1}$. Then

$$(309) \quad \xi^k = \cos kf + i \sin kf,$$

with similar formulas for ξ^{-k}, η^k, and η^{-k}.

Write the eccentricity as

$$(310) \quad e = \sin \varphi, \quad 0 < \varphi < \frac{1}{2}\pi.$$

For convenience, set

$$(311) \quad \beta = \tan \frac{1}{2}\varphi.$$

Then $0 < \beta < 1$, and e and β are related by the formulas

$$(312) \quad e = \frac{2\beta}{1 + \beta^2} \quad \text{and} \quad \beta = \frac{1 - \sqrt{1 - e^2}}{e}.$$

If e is small, $\beta \approx \frac{1}{2}e$. Furthermore,

$$(313) \quad \sqrt{\frac{1+e}{1-e}} = \frac{1+\beta}{1-\beta}.$$

We proved in Chapter II.3 that

(314) $$\tan \tfrac{1}{2}f = \sqrt{\frac{1+e}{1-e}}\,\tan \tfrac{1}{2}E = \frac{1+\beta}{1-\beta}\,\tan \tfrac{1}{2}E.$$

From the definition of ξ,

(315) $$i\tan \tfrac{1}{2}f = \frac{e^{if}-1}{e^{if}+1} = \frac{\xi-1}{\xi+1}.$$

Similarly,

(316) $$i\tan \tfrac{1}{2}E = \frac{\eta-1}{\eta+1}.$$

Hence, the relation between true and eccentric anomalies can be written

(317) $$\frac{\xi-1}{\xi+1} = \frac{1+\beta}{1-\beta}\frac{\eta-1}{\eta+1}.$$

In consequence,

(318) $$\xi = \frac{\eta-\beta}{1-\beta\eta} = \frac{\eta(1-\beta\eta^{-1})}{1-\beta\eta}$$

and

(319) $$\eta = \frac{\xi(1+\beta\xi^{-1})}{1+\beta\xi}.$$

(Notice that the relations between ξ and η interchange the roles of the variables upon interchange of β and $-\beta$.)

Now $\cos E = \tfrac{1}{2}(\eta+\eta^{-1})$; therefore

(320) $$r = a(1 - e\cos E) = a\left[1 - \frac{\beta}{1+\beta^2}(\eta+\eta^{-1})\right].$$

This gives

(321) $$\frac{r}{a} = \frac{(1-\beta\eta)(1-\beta\eta^{-1})}{1+\beta^2}.$$

From (169),

(322) $$\frac{r}{a} = \frac{1-e^2}{1+e\cos f} = \frac{(1-\beta^2)^2}{1+\beta^2}\frac{1}{(1+\beta\xi)(1+\beta\xi^{-1})}$$

in a similar manner.

The powers of β occur in some expansions, and it is convenient to have an expression for β^m in terms of e without intervention of the parameter φ. The equation (312) can be written in the form

$$(323) \qquad \beta = \frac{e}{2} + \frac{e}{2}\beta^2,$$

appropriate for application of Lagrange's Expansion Theorem. We use the apparatus as described in §2. Set $y = \beta$, $x = \frac{1}{2}e$, $\alpha = \frac{1}{2}e$, and $\varphi(x) = x^2$, with $F(y) = y^m$. Then

$$
(324) \qquad
\begin{aligned}
\beta^m &= x^m + \sum_{n=1}^{\infty} \frac{\alpha^n}{n!} \frac{d^{n-1}}{dx^{n-1}} \left[x^{2n} \frac{d}{dx} x^m \right] \\
&= x^m + m \sum_{n=1}^{\infty} \frac{\alpha^n}{n!} \frac{d^{n-1}}{dx^{n-1}} \left[x^{2n+m-1} \right] \\
&= x^m + m \sum_{n=1}^{\infty} \frac{\alpha^n}{n!} \frac{(2n+m-1)!}{(n+m)!} x^{n+m}.
\end{aligned}
$$

In terms of e, setting $\alpha = x = \frac{1}{2}e$,

$$
(325) \qquad
\begin{aligned}
\beta^m &= \left(\frac{e}{2}\right)^m \left\{ 1 + m \sum_{n=1}^{\infty} \frac{(2n+m-1)!}{n!\,(n+m)!} \left(\frac{e}{2}\right)^{2n} \right\} \\
&= \left(\frac{e}{2}\right)^m \left\{ 1 + \frac{m}{4}e^2 + \frac{m(m+3)}{4^2 \cdot 2!} e^4 \right. \\
&\qquad\qquad \left. + \frac{m(m+4)(m+5)}{4^3 \cdot 3!} e^6 + \cdots \right\}.
\end{aligned}
$$

6. Some algebraically-derived expansions

In this section we obtain Fourier expansions of a number of periodic functions of E and f. Having set $\eta = e^{iE}$, our procedure is to obtain Maclaurin or Laurent series in η and then convert these series into Fourier expansions.

From (318) we have

$$(326) \qquad \xi = \frac{\eta - \beta}{1 - \beta\eta} = \frac{\eta(1 - \beta\eta^{-1})}{1 - \beta\eta}.$$

Take logarithms:[7]

(327) $$if = iE + \log(1 - \beta\eta^{-1}) - \log(1 - \beta\eta).$$

We know that

(328) $$\log(1 - x) = -\sum_{n=1}^{\infty} \frac{x^n}{n} \qquad \text{for } |x| < 1.$$

Here, $|\eta| = 1$ and $|\beta| < 1$, so that $|\beta\eta| < 1$ and $|\beta\eta^{-1}| < 1$. Therefore,

(329)
$$
\begin{aligned}
if &= iE - \sum_{n=1}^{\infty} \frac{\beta^n \eta^{-n}}{n} + \sum_{n=1}^{\infty} \frac{\beta^n \eta^n}{n} \\
&= iE + \sum_{n=1}^{\infty} \frac{\beta^n}{n}(\eta^n - \eta^{-n}) \\
&= iE + 2i \sum_{n=1}^{\infty} \frac{\beta^n}{n} \sin nE.
\end{aligned}
$$

We conclude by taking imaginary parts that

(330) $$f = E + 2 \sum_{n=1}^{\infty} \frac{\beta^n}{n} \sin nE.$$

The expansion for E in terms of f is immediate from the note following equation (319): We need only interchange E and f and replace β by $-\beta$ to get

(331) $$E = f + 2 \sum_{n=1}^{\infty} (-1)^n \frac{\beta^n}{n} \sin nf.$$

As a function of E,

(332) $$r = a(1 - e \cos E);$$

from this relation, the Fourier expansion of r/a is obvious. The expansion of a/r is not obvious, except that only the cosine terms should appear in expansions of a/r in terms of E or of f. From (321), we make an

---

[7] Because we take logarithms of complex-valued quantities, we must be careful about logarithmic branches.

expression for a/r to which we apply the technique of partial fractions:

$$
\begin{aligned}
(333) \qquad \frac{a}{r} &= \frac{1 + \beta^2}{(1 - \beta\eta)(1 - \beta\eta^{-1})} \\
&= \frac{1 + \beta^2}{1 - \beta^2} \left\{ \frac{1}{1 - \beta\eta} + \frac{\beta}{\eta} \frac{1}{1 - \beta/\eta} \right\} \\
&= \frac{1 + \beta^2}{1 - \beta^2} \left\{ \sum_{n=0}^{\infty} \beta^n \eta^n + \sum_{n=0}^{\infty} \frac{\beta^{n+1}}{\eta^{n+1}} \right\} \\
&= \frac{1 + \beta^2}{1 - \beta^2} \left\{ 1 + \sum_{n=1}^{\infty} \beta^n (\eta^n + \eta^{-n}) \right\} \\
&= \frac{1}{\sqrt{1 - e^2}} + 2 \sum_{n=1}^{\infty} \frac{\beta^n}{\sqrt{1 - e^2}} \cos nE.
\end{aligned}
$$

Next,

$$
(334) \qquad \frac{a}{r} = \frac{1}{1 - e^2} + \frac{e}{1 - e^2} \cos f,
$$

a simple Fourier expansion. Furthermore,

$$
\begin{aligned}
(335) \qquad \frac{r}{a} &= \frac{(1 - \beta^2)^2}{1 + \beta^2} \frac{1}{(1 + \beta\xi)(1 + \beta\xi^{-1})} \\
&= \frac{1 - \beta^2}{1 + \beta^2} \left\{ \frac{1}{1 + \beta\xi} - \frac{\beta}{\xi} \frac{1}{1 + \beta/\xi} \right\} \\
&= \frac{1 - \beta^2}{1 + \beta^2} \left\{ \sum_{n=0}^{\infty} (-1)^n \beta^n \xi^n - \sum_{n=0}^{\infty} (-1)^n \frac{\beta^{n+1}}{\xi^{n+1}} \right\} \\
&= \frac{1 - \beta^2}{1 + \beta^2} \left\{ 1 + 2 \sum_{n=1}^{\infty} (-1)^n \beta^n \cos nf \right\}.
\end{aligned}
$$

EXERCISE IV.5. Expand $(r/a)^2$ and $(a/r)^2$ into Fourier series in E and in f. Partial answer:

$$
\begin{aligned}
\frac{a^2}{r^2} &= (1 - e^2)^{-3/2} \left[1 + 2 \sum_{n=1}^{\infty} \left\{ 1 + n\sqrt{1 - e^2} \right\} \beta^n \cos nE \right], \\
\frac{r^2}{a^2} &= \sqrt{1 - e^2} \left[1 + 2 \sum_{n=1}^{\infty} \left\{ 1 + n\sqrt{1 - e^2} \right\} (-\beta)^n \cos nf \right].
\end{aligned}
$$

(Expansions of the powers $(r/a)^p$ and $(a/r)^p$ can be worked out in general. The coefficients involve the so-called *hypergeometric functions*.)

It is possible to expand $r^p \cos qf$ and $r^p \sin qf$ in terms of E; these expansions include the results of Exercise IV.5 and its following remark, and they have various applications in the study of planetary motions. We consider the special cases $r^p \cos pf$ and $r^p \sin pf$; p need not be an integer.

We have

$$(336) \qquad \frac{r}{a} = \frac{(1 - \beta\eta)(1 - \beta\eta^{-1})}{1 + \beta^2}$$

and

$$(337) \qquad \xi = \frac{\eta(1 - \beta\eta^{-1})}{1 - \beta\eta}.$$

Therefore,

$$(338) \qquad r\xi = \frac{a\eta(1 - \beta\eta^{-1})^2}{1 + \beta^2},$$

and

$$(339) \qquad r^p\xi^p = a^p(1 + \beta^2)^{-p}\eta^p(1 - \beta\eta^{-1})^{2p}.$$

The binomial coefficient $\binom{p}{n}$ is defined for integer $n \geq 0$ and arbitrary p by the binomial expansion

$$(340) \qquad (1 + x)^p = \sum_{n=0}^{p} \binom{p}{n} x^n, \quad |x| < 1,$$

in which

$$(341) \qquad \binom{p}{n} = \frac{p(p - 1)(p - 2) \cdots (p - n + 1)}{n!}.$$

Therefore,

$$(342) \qquad \begin{aligned} r^p\xi^p &= a^p(1 + \beta^2)^{-p}\eta^p \sum_{n=0}^{\infty}(-1)^n \binom{p}{n}\beta^n\eta^{-n} \\ &= a^p(1 + \beta^2)^{-p} \sum_{n=0}^{\infty}(-1)^n \binom{p}{n}\beta^n\eta^{p-n}. \end{aligned}$$

Taking real and imaginary parts, we obtain

$$(343) \qquad r^p \cos pf = a^p(1 + \beta^2)^{-p} \sum_{n=0}^{\infty}(-\beta)^n \binom{p}{n} \cos{(p - n)}E$$

and

$$(344) \qquad r^p \sin pf = a^p (1 + \beta^2)^{-p} \sum_{n=0}^{\infty} (-\beta)^n \binom{p}{n} \sin (p - n) E.$$

These expansions do not resemble standard Fourier series if p is not an integer, but they can be put into the standard form (294) after the sine and cosine are expanded by their addition theorems. If p is an integer, then the series are clearly Fourier series, and they have finitely many summands, terminating with the terms in which $n = p$.

The functions $\cos f$ and $\sin f$ have expansions in terms of E. We have from (334) that

$$(345) \qquad \begin{aligned} \cos f &= \frac{1 - e^2}{e} \frac{a}{r} - \frac{1}{e} \\ &= \frac{1 - e^2}{e} \frac{1}{\sqrt{1 - e^2}} \left\{ 1 + 2 \sum_{n=1}^{\infty} \beta^n \cos nE \right\} - \frac{1}{e} \\ &= -\frac{1 - \sqrt{1 - e^2}}{e} + \frac{2\sqrt{1 - e^2}}{e} \sum_{n=1}^{\infty} \beta^n \cos nE, \end{aligned}$$

which can be put into the form

$$(346) \qquad \cos f = -\beta + (1 - \beta^2) \sum_{n=1}^{\infty} \beta^{n-1} \cos nE.$$

Because the last equation is obtained by taking the real part of

$$(347) \qquad \xi = -\beta + (1 - \beta^2) \sum_{n=1}^{\infty} \beta^{n-1} \eta^n,$$

we may also take the imaginary part to obtain

$$(348) \qquad \sin f = (1 - \beta^2) \sum_{n=1}^{\infty} \beta^{n-1} \sin nE.$$

By using the interchange noted after (319), we can immediately write

$$(349) \qquad \cos E = \beta + (1 - \beta^2) \sum_{n=1}^{\infty} (-\beta)^{n-1} \cos nf$$

and

$$(350) \qquad \sin E = (1 - \beta^2) \sum_{n=1}^{\infty} (-\beta)^{n-1} \sin nf.$$

We can express the mean anomaly M in terms of f. Starting from

(351)
$$\frac{a}{r} = \frac{1 + e \cos f}{1 - e^2},$$

we find that

(352)
$$\frac{dr}{df} = \frac{er^2 \sin f}{a(1 - e^2)}.$$

Using (175), we have

(353)
$$r \sin f = b \sin E = a\sqrt{1 - e^2} \sin E.$$

Hence

(354)
$$\frac{dr}{df} = \frac{er \sin E}{\sqrt{1 - e^2}}.$$

We know that $r = a(1 - e \cos E)$. Kepler's equation tells us that $E - \sin E = M$. Therefore,

(355)
$$\frac{dr}{dM} = ae \sin E \frac{dE}{dM} = \frac{a^2 e}{r} \sin E.$$

By the chain rule and the result of Exercise IV.5,

(356)
$$\frac{dM}{df} = \frac{r^2}{a^2\sqrt{1 - e^2}}$$
$$= 1 + 2 \sum_{n=1}^{\infty} (-1)^n \{1 + n\sqrt{1 - e^2}\} \beta^n \cos nf.$$

Remembering that M and f vanish together, we can integrate to get

(357)
$$M = f + 2 \sum_{n=1}^{\infty} (-1)^n \left\{ \frac{1}{n} + \sqrt{1 - e^2} \right\} \beta^n \sin nf.$$

The angle $f - M$ is called the *equation of the center* (see (183)), and the last series gives its expansion in terms of the true anomaly.

EXERCISE IV.6. Evaluate these integrals as functions of n (an integer) and e:

(a) $\displaystyle\int_0^{2\pi} \frac{\cos nE \, dE}{1 - e \cos E}.$

(b) $\displaystyle\int_0^{2\pi} \frac{\cos nE \, dE}{(1 - e \cos E)^2}.$

(c) $\displaystyle\int_0^{2\pi} \frac{\sin nE \, dE}{1 - e \cos E}.$

(d) $\displaystyle\int_0^{2\pi} \frac{\sin nE\, dE}{(1 - e\cos E)^2}.$

(e) $\displaystyle\int_0^{2\pi} \cos\left\{ 2\tan^{-1}\left[\sqrt{\frac{1+e}{1-e}}\, \tan \tfrac{1}{2}E \right] \right\} \cos nE\, dE.$

(f) $\displaystyle\int_0^{2\pi} \sin\left\{ 2\tan^{-1}\left[\sqrt{\frac{1+e}{1-e}}\, \tan \tfrac{1}{2}E \right] \right\} \sin nE\, dE.$

EXERCISE IV.7. A particle describes an ellipse under an inverse-square law of attraction to a focus. Find, in terms of the eccentricity of the ellipse, the ratio of the time spent near the pericentron between the ends of the latus rectum to the total period of the orbit.

7. Expansions in terms of the mean anomaly

Kepler's equation relates the mean anomaly M and the eccentric anomaly E through the eccentricity e:

(358) $$E - e\sin E = M.$$

We discussed in II.5 methods for determining E in terms of M. Here we express E as a Fourier series with respect to M.

We have

(359) $$\frac{dM}{dE} = 1 - e\cos E,$$

so that

(360) $$\frac{dE}{dM} = \frac{1}{1 - e\cos E}.$$

We already know how to expand the last fraction into a Fourier series in E (compare (333)), but we want the expansion in terms of M. Once we determine the expansion of dE/dM, we can get that of E itself by an integration.

Note that E is an odd function of M, so that dE/dM is an even function of M. Therefore, dE/dM is expandable into a Fourier cosine series:

(361) $$\frac{dE}{dM} = \tfrac{1}{2}a_0 + \sum_{n=1}^{\infty} a_n \cos nM.$$

Invoking evenness, the coefficient

$$(362) \quad a_n = \frac{1}{\pi} \int_{-\pi}^{\pi} \frac{dE}{dM} \cos nM \, dM = \frac{2}{\pi} \int_{0}^{\pi} \frac{dE}{dM} \cos nM \, dM.$$

Because M and E pass through 0 and through π together, we use Kepler's equation along with the chain rule to get

$$
(363) \qquad
\begin{aligned}
a_n &= \frac{2}{\pi} \int_{0}^{\pi} \cos[n(E - e \sin E)] \, dE \\
&= \frac{2}{\pi} \int_{0}^{\pi} \cos(nE - e \sin E) \, dE \\
&= 2 J_n(ne),
\end{aligned}
$$

by Bessel's integral (293). Therefore,

$$(364) \qquad \frac{dE}{dM} = 1 + 2 \sum_{n=1}^{\infty} J_n(ne) \cos nM.$$

By integrating, we get

$$(365) \qquad E = M + 2 \sum_{n=1}^{\infty} \frac{1}{n} J_n(ne) \sin nM.$$

(The name of Bessel is commonly associated to this expansion, but it is evidently due to Lagrange.)

Now we derive expansions of some functions of f, M, and E.

Write

$$(366) \qquad \cos kE = \tfrac{1}{2} a_{k,0} + \sum_{n=1}^{\infty} a_{k,n} \cos nM.$$

Then, exploiting evenness and integrating by parts to introduce dE/dM,

$$
\begin{aligned}
a_{k,n} &= \frac{2}{\pi} \int_0^\pi \cos kE \cos nM \, dM \\
(367) \qquad &= \frac{2}{n\pi} [\cos kE \sin nM]_0^\pi + \frac{2k}{n\pi} \int_0^\pi \sin kE \sin nM \, dE \\
&= \frac{k}{n\pi} \int_0^\pi \cos(nM - kE) \, dE - \frac{k}{n\pi} \int_0^\pi \cos(nM + kE) \, dE \\
&= \frac{k}{n\pi} \int_0^\pi \cos[(n - k)E - ne \sin E] \, dE \\
&\qquad - \frac{k}{n\pi} \int_0^\pi \cos[(n + k)E - ne \sin E] \, dE \\
&= \frac{k}{n} [J_{n-k}(ne) - J_{n+k}(ne)].
\end{aligned}
$$

This formula is meaningless if $n = 0$, but

$$
\begin{aligned}
a_{k,0} &= \frac{2}{\pi} \int_0^\pi \cos kE \, (1 - e \cos E) \, dE \\
&= \frac{2}{\pi} \int_0^\pi \cos kE \, dE - \frac{e}{\pi} \int_0^\pi \cos (k + 1)E \, dE \\
&\qquad - \frac{e}{\pi} \int_0^\pi \cos (k - 1)E \, dE \\
&= \begin{cases} 0, & \text{if } k > 1, \\ -e, & \text{if } k = 1. \end{cases}
\end{aligned}
$$

Hence,

$$
(368) \qquad \cos E = -\tfrac{1}{2}e + \sum_{n=1}^\infty \frac{1}{n} [J_{n-1}(ne) - J_{n+1}(ne)] \cos nM,
$$

and, by (284), this can be written as

$$
(369) \qquad \cos E = -\tfrac{1}{2}e + \sum_{n=1}^\infty \frac{2}{n^2} \frac{d}{de} J_n(ne) \cdot \cos nM.
$$

If $k > 1$, then

$$
(370) \qquad \cos kE = k \sum_{n=1}^\infty \frac{1}{n} [J_{n-k}(ne) - J_{n+k}(ne)] \cos nM.
$$

By similar calculations,

$$(371) \qquad \sin E = \sum_{n=1}^{\infty} \frac{1}{n}[J_{n+1}(ne) + J_{n-1}(ne)] \sin nM$$

$$= \frac{2}{e} \sum_{n=1}^{\infty} \frac{1}{n} J_n(ne) \sin nM;$$

if $k > 1$, then

$$(372) \qquad \sin kE = k \sum_{n=1}^{\infty} \frac{1}{n}[J_{n+k}(ne) + J_{n-k}(ne)] \sin nM.$$

From Kepler's equation,

$$(373) \qquad \frac{dE}{dM} = \frac{a}{r}.$$

The expansion (364) of dE/dM gives

$$(374) \qquad \frac{a}{r} = 1 + 2 \sum_{n=1}^{\infty} J_n(ne) \cos nM.$$

We have, using Kepler's equation,

$$\frac{d}{dM}\left(\frac{r^2}{a^2}\right) = \frac{2r}{a^2}\frac{dr}{dE}\frac{dE}{dM} = 2\sin E$$

$$= 4 \sum_{n=1}^{\infty} \frac{1}{n} J_n(ne) \sin nM.$$

We may integrate term by term:[8] There is a constant C such that

$$(375) \qquad \frac{r^2}{a^2} = C - 4 \sum_{n=1}^{\infty} \frac{1}{n^2} J_n(ne) \cos nM.$$

Now

$$(376) \quad \frac{r^2}{a^2} = (1 - e\cos E)^2 = 1 + \tfrac{1}{2}e^2 - 2e\cos E + \tfrac{1}{2}e^2 \cos 2E.$$

From the expansions (369) of $\cos E$ and (370) of $\cos kE$, the nonperiodic part of r^2/a^2 is $1 + \tfrac{1}{2}e^2 - 2e(-\tfrac{1}{2}e) = 1 + \tfrac{1}{2}e^2$. Consequently,

$$(377) \qquad \frac{r^2}{a^2} = 1 + \tfrac{3}{2}e^2 - 4 \sum_{n=1}^{\infty} \frac{1}{n^2} J_n(ne) \cos nM.$$

---

[8] Whether a Fourier series converges or not, the series found by integrating term by term always converges to the integral of the function expanded. E.C.Titchmarsh, *Theory of Functions*, §13.5.

Next,

(378)
$$e \cos f = -1 + (1 - e^2)\frac{a}{r}$$
$$= -1 + (1 - e^2)\left[1 + 2\sum_{n=1}^{\infty} J_n(ne) \cos nM\right],$$

so that

(379)
$$\cos f = -e + \frac{2(1 - e^2)}{e} \sum_{n=1}^{\infty} J_n(ne) \cos nM.$$

Also,

(380)
$$r \sin f = a\sqrt{1 - e^2} \sin E = \frac{\sqrt{1 - e^2}}{e} \frac{dr}{dE}.$$

But

(381)
$$\frac{dr}{dE} = \frac{dr}{dM}\frac{dM}{dE} = \frac{r}{a}\frac{dr}{dM},$$

so that

(382)
$$\sin f = \frac{\sqrt{1 - e^2}}{e} \frac{d}{dM}\left(\frac{r}{a}\right).$$

Combining the relation $r/a = 1 - e \cos E$ with the expansion (369), we find the expansion

(383)
$$\sin f = 2\sqrt{1 - e^2} \sum_{n=1}^{\infty} \frac{1}{n}\frac{d}{de} J_n(ne) \sin nM.$$

Finally, we obtain expansions for the components along the major and minor axes of the attraction toward the focus; precisely, we expand

(384)
$$\frac{\cos f}{r^2} \quad \text{and} \quad \frac{\sin f}{r^2}.$$

Let (ξ, η) be the rectangular coordinates of the planet in its orbit. Then, with the focus as origin, we have (175)

(385)
$$\frac{\xi}{a} = \cos E - e \quad \text{and} \quad \frac{\eta}{a} = \sqrt{1 - e^2} \sin E.$$

From the expansions (369) and (371), we find that

$$(386) \qquad \frac{\xi}{a} = -\tfrac{3}{2}e + 2\sum_{n=1}^{\infty} \frac{1}{n^2}\frac{d}{de}J_n(ne)\cos nM,$$

$$(387) \qquad \frac{\eta}{a} = \frac{2}{e}\sqrt{1-e^2}\sum_{n=1}^{\infty}\frac{1}{n}J_n(ne)\sin nM.$$

But $\xi = r\cos f$ and $\eta = r\sin f$; therefore,

$$(388) \qquad \frac{\xi}{r^3} = \frac{\cos f}{r^2} \quad \text{and} \quad \frac{\eta}{r^3} = \frac{\sin f}{r^2}.$$

The equations of motion in rectangular coordinates are

$$(389) \qquad \ddot{\xi} + \frac{\mu\xi}{r^3} = 0 \quad \text{and} \quad \ddot{\eta} + \frac{\mu\eta}{r^3} = 0.$$

Because $M = n(t-\tau) = (\mu a^{-3})^{1/2}(t-\tau)$, $d/dt = n\,d/dM$ and $d^2/dt^2 = n^2\,d^2/dM^2 = \mu/a^3\,d^2/dM^2$. Then

$$(390) \qquad \frac{d^2\xi}{dM^2} + \frac{a^3\xi}{r^3} = 0 \quad \text{and} \quad \frac{d^2\eta}{dM^2} + \frac{a^3\eta}{r^3} = 0.$$

Hence, we can get expansions for $\cos f/r^2 = \xi/r^3$ and $\sin f/r^2 = \eta/r^3$ by differentiating the expansions for ξ/a and η/a. The results are

$$(391) \qquad \frac{\cos f}{r^2} = \frac{2}{a^2}\sum_{n=1}^{\infty}\frac{d}{de}J_n(ne)\cos nM$$

$$(392) \qquad \frac{\sin f}{r^2} = \frac{2\sqrt{1-e^2}}{a^2 e}\sum_{n=1}^{\infty} nJ_n(ne)\sin nM.$$

EXERCISE IV.8. The *cycloid* is the curve traced by a point on the circumference of a circular disk as the disk rolls without slipping along a straight line.

(a) If the line is the x-axis, the disk has radius 1, and the disk rolls above the line, then show that the cycloid has parametric equation

$$(393) \qquad x = \vartheta - \sin\vartheta \quad \text{and} \quad y = 1 - \cos\vartheta,$$

where the point begins at the origin and ϑ is the angle through which the disk has turned.

(b) Clearly y is a periodic function of x. Express y as a Fourier series in terms of x. [Higher derivatives of the Bessel functions J_n will be needed because it will be necessary to differentiate Bessel's integral with respect to the parameters in it.]

Gravitation and Closed Orbits

1. Bertrand's characterization of universal gravitation

The solar system appears to be stable: Calculations into the future using Newton's Law of Universal Gravitation show that the Sun and the planets will behave in the future much as they have done in the past. The Earth will move in its yearly, very nearly elliptical, Keplerian orbit around the Sun, as will the other planets in their years. No planet will suddenly take off for deep space or fall into the Sun, nor will two of the (major) planets collide.

Of course, there are minor flaws in this rosy description. The perihelion of the planet Mercury is observed to rotate slightly ahead with each orbit around the Sun. While this advance could be accounted for if the Sun were oblate, careful observations of the Sun have yielded a maximum possible oblateness that is wholly inadequate to account for the advance. General Relativity Theory correctly predicts the observed advance for Mercury and predicts that the same effect for the other planets is unobservably small. Planetary bodies do collide, although the collisions we see are mostly meteors colliding with the Earth, and—a few notable exceptions ignored—with trivial consequences. There are some significant interactions. For example, the time of perihelion for Halley's Comet in 1986 was altered because of a close pass by Jupiter. Comet Shoemaker-Levy 9 was captured by Jupiter as it passed by in 1993 and broken up, and the score or so of fragments returned to plunge into Jupiter's atmosphere in July, 1994. The existence of the planets Neptune and Pluto was predicted to account for observed 'irregularities' in the calculated motions for the planets Uranus and Neptune, respectively.

However, what has always been evident about the solar system is its regularity. The motions are so regular that they can be well represented by a clockwork model, called an orrery.[1] To a very satisfactory degree, the planets move independently of one another, as if each were the only traveler around the Sun in a solitary Kepler ellipse. How much of this regularity is a result of Newton's Law of Universal Gravitation? We have seen (p. 33) that possible orbits for the inverse cube power law include the Cotes spirals, which are neither closed nor even bounded. What power laws result in orbits closed when they are bounded? A complete answer was given by J. Bertrand in 1873:

> The only central power laws of force for which every bounded orbit is closed are the inverse square law (exponent -2) and Hooke's law (exponent 1).

Only the inverse square law gives forces that die off as distances increase, so Newton's Law of Universal Gravitation is the only power law that can account for the observed regularity of the solar system. Hooke's Law gives elliptical orbits but with the center of force at the center of the ellipse, and the force toward the center increases with distance.

We will prove Bertrand's Theorem in this chapter.[2] We already know that the bounded orbits in the cases of Newton's and Hooke's Laws are

---

[1] Harriet Wvnter and Anthony Turner, *Scientific Instruments,* Scribner's, New York, 1975, pp. 43–50. The orrery is named ror Charles Boyle, fourth Earl of Orrery, for whom the *second* example was made by John Rowley in the early eighteenth century. It reproduced the daily motion of the Earth and the periods of the moon, and so should have been called a 'tellurium,' but Rowley's honorific persists. The name 'planetarium' was also used but now is taken to refer to a device for projecting points of light onto a hemispherical dome.

[2] We learned of this beautiful theorem by reading in H. Goldstein, *Classical Mechanics, Second Edition,* Addison-Wesley, 1980. We regret that it was necessary to strip away the elegant development into which Goldstein embedded it in order to present it in the context of this book, and we recommend reading Goldstein for its mathematical comprehensiveness and its insight.

ellipses, and so are closed. Therefore, it will be enough to rule out the other exponents by proving that each of the others admits *at least one* bounded orbit that is not closed.

EXERCISE V.1. Which power laws admit unbounded radial motions?

2. Circular motions

We begin by examining circular orbits. We have derived the equations (83) and, in integrated form, (85) for the planar motion of a mass point under a central force of magnitude $P(r)$. Suppose that the point moves in a circle of *constant* radius r centered at the center of force. Then $\ddot{r} = \dot{r} = 0$, so that the equations become

$$\text{(394)} \qquad r\dot{\vartheta}^2 = P(r) \quad \text{and} \quad r^2\dot{\vartheta} = h.$$

The second of these equations shows that we may proceed in terms of either $\omega = \dot{\vartheta}$ or h. We find that

$$\text{(395)} \qquad \omega^2 = P(r)/r \quad \text{and} \quad h^2 = r^3 P(r).$$

In particular, there can exist a circular motion only at those radii for which $P(r) > 0$. Given an admissible r, we find that there is a motion with angular velocity $\omega = \sqrt{P(r)/r}$ (or its negative).

Whether we can preassign the angular velocity and then obtain the radius (or the radii) from (395) depends upon the nature of the function $P(r)$.

EXERCISE V.2. In the movie 'Journey to the Far Side of the Sun,' an astronaut blasts off from Earth to begin a journey to Venus. An explosion occurs soon after liftoff, and the astronaut wakes up in a hospital. Soon he begins to notice disturbing differences in chirality: most people are left-handed and wear wedding rings on the right hand, coats button on the 'wrong' side, and so on. He learns that he is on Anti-Earth, which is the same mass and shape as the Earth and which revolves around the Sun in the same orbit but always on the other end of the Earth's diameter. Anti-Earth can not be seen through Earth-based telescopes. Anti-Earth does know about Earth and—naturally—is hostile to it. An annihilating invasion is about to be launched. The astronaut manages to escape, with the help of a pretty nurse, and they get back to warn Earth.

Use the following plan to show that the existence of an Anti-Earth could have (and would have) been deduced long ago. Suppose that the Earth is represented as a point of mass m revolving in a circular orbit of radius a around a center of force (the Sun) of mass M. Increase the usual Newtonian attraction of the Sun by the attraction of the hidden Anti-Earth and compute the change in the length of the Earth year under the new central force. (How many seconds longer or shorter will the year be under the Anti-Earthian influence than the 'real' year is?)

The presence of the Anti-Earth would be seen from the Earth as an increase in the constant GM, and that changed constant would then change the periods of the other planets when calculated from Kepler's Third Law. But those periods are observed to be in harmony with the period of the Earth when it is calculated from Kepler's Third Law. Conclusion: no Anti-Earth.

EXERCISE V.3. (a) Suppose that a 'planet' of mass m moves in a circle of radius a around a 'sun' of mass M. Show that a massless planetoid can revolve in a circular orbit around the sun and always remain on the line through the sun and the planet. (There are three possible positions. You will get a polynomial of degree 5 to solve, but two of its roots are complex. This is the Lagrangian straight-line solution for a simple, planar version of the so-called 'restricted three body problem.' The masslessness of the third body means that, while the other two bodies act on it, it does not affect the motion of the other two bodies.)

(b) Use numerical methods to find the possible radii as multiples of a for the Earth-Moon system.

3. Neighbors of circular motions

We can reformulate the discussion of the previous section in terms of potential functions. Suppose that $V(r)$ is a potential function for the force per unit mass of magnitude $-P(r)$. Then we have seen that the three- (or two-)dimensional motion under a central force in a system rotating with constant angular velocity $\omega = \dot{\vartheta}$ can be expressed in terms of a one-dimensional motion along the r-axis with modified potential $\tilde{V} = V - \frac{1}{2}r^2\omega^2$. The condition (395) that we have found to characterize circular orbits can be rewritten simply as $\tilde{V}'(r) = 0$, which is to say that the radius r and the angular velocity ω must be such as to produce a critical value for $\tilde{V}$. In yet another reformulation, we observe that r remains stationary whenever ω is such as to make the *modified force* $-\nabla\tilde{V} = 0$.

Suppose that a 'particle' is traveling in a circular orbit when it is slightly perturbed. For example, a rocket traveling in a circular orbit might be moved into a new orbit by a short blast of its engines, it might collide with a small bit of celestial debris, or it might set out and leave an artificial satellite. If the magnitude of the cause is small, then the magnitude of the effect may be small. Because the particle is moving over an interval of time, we can sort out the effects as viewed over time according to whether the effects remain small and bounded or become large. In the first case, we say that the circular orbit is *stable* and in the other case, *unstable*.[3] In the rest of this section, we will take a first look at the radial stability of circular orbits; that is, at the question of stability when the orbit is altered slightly in the radial direction.

Because the $r\vartheta$-equation (88) is highly nonlinear and we want to keep down tedious algebra, we will work with the transformed $u\vartheta$-equation (89):

$$(396) \qquad \frac{d^2u}{d\vartheta^2} + u = \frac{P(1/u)}{h^2 u^2}.$$

If we specialize to the power rule $P(r) = \mu/r^\beta$, then (396) becomes

$$(397) \qquad \frac{d^2u}{d\vartheta^2} + u = \frac{\mu}{h^2} u^{\beta-2}.$$

Suppose that a circular orbit has radius $r_0 = 1/u_0$, angular momentum h, and angular velocity ω, these quantities being related by (394). Then $u_0^{\beta-3}\mu = h^2$. There is no loss in generality if we assume that $u_0 = 1$, because this can be assured merely by adjusting the unit of length. Consequently, $\mu/h^2 = 1$. Our algebraic calculations will now be simpler.

We will slightly perturb the motion and examine the results. Let ϵ be a 'small' parameter whose square may be neglected. Write $u(\vartheta) = u_0 + \epsilon u_1(\vartheta)$, where $u_1(\vartheta)$ is a function of 'ordinary' magnitude. In fact, our specification of u_0 means that $u(\vartheta) = u_0 + \epsilon u_1(\vartheta)$. Neglecting ϵ^2 and higher powers, we find that

$$(398) \qquad u^{\beta-2} = (1 + \epsilon u_1)^{\beta-2} = 1 + (\beta - 2)\epsilon u_1.$$

Inserting this expansion into the differential equation (397), we get

$$(399) \qquad 1 + \epsilon \left(\frac{d^2u_1}{d\vartheta^2} + u_1 \right) = 1 + (\beta - 2)\epsilon u_1.$$

---

[3]This is a very crude sorting. There are many subtle refinements of the notion of stability. Over fifty criteria for stability are described and compared in V. Szebehely, 'Review of Concepts of Stability,' *Celestial Mechanics*, 64(1984), 49–64.

This is to be true for all 'small' ϵ, so the coefficients of corresponding powers of ϵ must be equal. Equating the coefficients of ϵ^0 gives only $1 = 1$, no surprise. The coefficients of ϵ^1 lead to the relation

$$(400) \qquad \frac{d^2 u_1}{d\vartheta^2} + u_1 = (\beta - 2)\epsilon u_1,$$

which simplifies to

$$(401) \qquad \frac{d^2 u_1}{d\vartheta^2} + (3 - \beta)u_1 = 0.$$

Because of the circular symmetry of the unperturbed orbit, we may choose the direction of the polar coordinate axis at will. This is the same as choosing one of the two arbitrary constants in the general solution of the differential equation. The differential equation (401) will have only bounded solutions exactly when $3 - \beta > 0$. Setting $\lambda = \sqrt{3 - \beta} > 0$, we take $u_1(\vartheta) = A \cos \lambda\vartheta$, where A is the second and remaining arbitrary constant.

We now have a first reduction: For every bounded orbit of an $r^{-\beta}$ law to be closed, it is necessary that $\beta < 3$. But there is more. In our context, the adjective 'closed' is synonymous with 'periodic.' The orbit must begin to retrace itself after finitely many laps around the center of force. This means that only *rational* values of λ are admissible, and we can write $\lambda = p/q$, a rational number in lowest terms. In other words, $\beta = 3 - (p/q)^2$.

EXERCISE V.4. Where did the quantity ϵ come from?

EXERCISE V.5. (a) An infinitely long, thin wire of linear density ρ runs along the z-axis. Show that the attraction $P(r)$ of the wire on a point in the xy-plane is proportional to $1/r$, with $r = \sqrt{x^2 + y^2}$.

(b) Is every bounded orbit closed for the $1/r$ attraction?

4. Higher perturbations; completion of the proof

We have eliminated infinitely many possibilities for β, but what remains is still an infinity—albeit a smaller one—of possible exponents. We now will reduce the possibilities to two.

Add two more terms to the perturbation expansion $u(\vartheta) = 1 + \epsilon u_1(\vartheta)$, setting

$$(402) \qquad u(\vartheta) = 1 + \epsilon u_1(\vartheta) + \epsilon^2 u_2(\vartheta) + \epsilon^3 u_3(\vartheta),$$

in which $u_2(\vartheta)$ and $u_3(\vartheta)$ are again functions of 'ordinary' magnitude. We will neglect ϵ^4 and higher powers.[4] After some tedious but routine algebra with the binomial series, we find that

$$
\begin{aligned}
u^{\beta-2} &= [1 + \epsilon u_1(\vartheta) + \epsilon^2 u_2(\vartheta) + \epsilon^3 u_3(\vartheta)]^{\beta-2} \\
(403) \quad &= 1 + \epsilon(\beta - 2)u_1 + \epsilon^2[(\beta - 2)u_2 + \tfrac{1}{2}(\beta - 2)(\beta - 3)u_1^2] \\
&\quad + \epsilon^3[(\beta - 2)u_3 + (\beta - 2)(\beta - 3)u_1 u_2 \\
&\quad\quad + \tfrac{1}{6}(\beta - 2)(\beta - 3)(\beta - 4)u_1^3].
\end{aligned}
$$

Now return to (397), filling in the left-hand side to match the expansion that we have derived for the right-hand side.[5] The coefficients of ϵ^0 and ϵ^1 lead, of course, to the equations that we have considered in the previous section. We consider the ϵ^2 and ϵ^3 coefficients.

First, the ϵ^2 coefficients. When we set $u_1 = A \cos \lambda\vartheta$ and carry out some algebra, we arrive at the differential equation

$$
\begin{aligned}
(404) \quad \frac{d^2 u_2}{d\vartheta^2} + \lambda^2 u_2 &= \tfrac{1}{2}(\beta - 2)(\beta - 3)A^2 \cos^2 \lambda\vartheta \\
&= \tfrac{1}{4}(\beta - 2)(\beta - 3)A^2(1 + \cos 2\lambda\vartheta),
\end{aligned}
$$

which has the solutions

$$(405)$$
$$u_2(\vartheta) = B \cos \lambda\vartheta + C \sin \lambda\vartheta - \tfrac{1}{4}(\beta - 2)A^2 + \tfrac{1}{12}(\beta - 2)A^2 \cos 2\lambda\vartheta.$$

This function also has $2\pi/\lambda$ as period, the same as does $u_1(\vartheta)$. We can draw no conclusions beyond those that we have already drawn from the earlier terms.

---

[4]Why not stop at ϵ^2 or go on to ϵ^4? The answer is simple, but revealing: In blocking out the calculation, we found that ϵ^2 brought no progress, but that adjoining ϵ^3 was sufficient. The description that you are reading is the cleaned-up, public version of a messy experiment.

This expansion consists of the first four terms of what—presumably—is a full Maclaurin series in ϵ. Does the series converge? *We do not care!* Only finitely many terms will be used in any calculation. There is a distinct algebraic—not analytic—flavor to the calculation.

[5]Remember that we have adjusted the unit of length with the consequence that $\mu/h^2 = 1$.

The ϵ^3 coefficients do yield more—and sufficient—information. When we substitute the expressions already found for $u_1(\vartheta)$ and $u_2(\vartheta)$ and carry out the necessary algebra, we obtain the differential equation

$$
\begin{aligned}
\frac{d^2 u_3}{d\vartheta^2} + \lambda^2 u_3 \;=\; & (\beta - 2)(\beta - 3)A\cos\lambda\vartheta \\
(406) \qquad & \times[\tfrac{1}{4}(2-\beta)A^2 + B\cos\lambda\vartheta + C\sin\lambda\vartheta \\
& \qquad + \tfrac{1}{12}(\beta - 2)A^2\cos 2\lambda\vartheta] \\
& + \tfrac{1}{6}(\beta - 2)(\beta - 3)(\beta - 4)A^3\cos^3\lambda\vartheta.
\end{aligned}
$$

The solution of this equation by formula is elementary, but we will not need the formula in full. We use trigonometric identities to reduce the right-hand side to a sum of sines and cosines of multiple angles. Forcing terms of the form $\cos k\lambda\vartheta$ or $\sin k\lambda\vartheta$ contribute only bounded summands to u_3 if $k \neq 1$. By the principle of superposition, we need look only at the product $\cos\lambda\vartheta\cos 2\lambda\vartheta = \tfrac{1}{2}[\cos\lambda\vartheta + \cos 3\lambda\vartheta]$ and the power $\cos^3\lambda\vartheta = \tfrac{1}{4}\cos 3\lambda\vartheta + \tfrac{3}{4}\cos\lambda\vartheta$. The presence of $\cos\lambda\vartheta$ on the right side of the differential equation will result in a term proportional to $\vartheta\cos\lambda\vartheta$ in the solution, and this function is unbounded. To ensure that $\cos\lambda\vartheta$ disappears from the right side, we demand that the coefficient of $\cos\lambda\vartheta$ vanish, which results in the equation

$$
\begin{aligned}
(407) \qquad & -\tfrac{1}{4}(\beta - 2)^2(\beta - 3) + \tfrac{1}{24}(\beta - 2)^2(\beta - 3) \\
& + \tfrac{1}{8}(\beta - 2)(\beta - 3)(\beta - 4) = 0.
\end{aligned}
$$

The roots of this equation are $\beta = -1, 2, 3$. However, the root $\beta = 3$ is already excluded by the condition $\beta < 3$ that came from the ϵ^1-term, leaving only $\beta = -1$ and $\beta = 2$. These two exponents do fulfill the conclusion of Bertrand's Theorem, as we have shown in earlier chapters.

EXERCISE V.6. Suppose that every bounded orbit of a certain central force with magnitude function $P(r)$ closes up after k turns around the center (which might include closing up after d turns, where d divides k). What can you say about $P(r)$? Suggestion: Use Bertrand's Theorem and the formalism of Exercise II.6, considering $U(\vartheta) = u(k\vartheta)$.

Having regard to our earlier mention of Hooke's Law, we arrive at a characterization of Newton's Law of Universal Gravitation:

Newton's Law of Universal Gravitation is the only power law that dies off as distances from the center of force increase and for which every bounded orbit is closed.

5. From differential geometry in the large

In this section we will develop some paraphernalia and prove a theorem from the differential geometry in the large of plane curves.[6] In the next section, we will translate that theorem to find a necessary condition for there to be a closed orbit of a certain type in an attracting central force field.

A continuous plane curve $r(t)$, $a \leq t \leq b$, is called *closed* if $r(a) = r(b)$. Alternatively, we can call $r(t)$, $-\infty < t < \infty$, *closed* if it is periodic; that is, if there is a number $p > 0$ such that $r(t + p) = r(t)$ for all t. The two definitions are in harmony: Set $p = b - a$ and extend $r(t)$ from $a \leq t \leq b$ to $-\infty < t < \infty$ by requiring that $r(t + p) = r(t)$ for all t. A closed curve is called *simple* if $r(t_1) = r(t_2)$ exactly when $t_1 - t_2 = kp$ for some integer k.[7] The curve is of *class C^2* if the vector function $r(t)$ has continuous second derivatives for all t, and it is *regular* if $dr/dt \neq 0$ for all t.

We can attach *Frenet apparatus* to a plane curve of class C^2 in the following way. Let s be an arc-length parameter for the curve so that $ds/dt = |dr/dt|$. Then $|dr/ds| = 1$ for all s. Define a vector field $T(s)$ along the curve by setting $T = dr/ds$, and note that $|T| = 1$. (See Figure V.1.) A second vector field along the curve is called N and is obtained by rotating $T(s)$ through a right angle in the counterclockwise sense. The vectors $T(s)$ and $N(s)$ are respectively the *(unit) tangent vector* and the

---

[6]We are led by D. Laugwitz, *Differential and Riemannian Geometry*, Academic Press, 1965, §16.1. Don't let the fraktur symbols and the continental vector notation which survived the translation from German to English scare you off—this is a concise, readable, yet wonderfully informative introduction to modern differential geometry.

[7]The *Jordan Curve Theorem*, the fundamental theorem about simple closed plane curves, states:

The image of a simple closed curve is a compact subset of the plane whose complement consists of two connected components, exactly one of which is bounded.

The bounded component is called the *interior* and the unbounded component, the *exterior*, of the curve.

(unit) normal vector to the curve at the point $r(s)$.

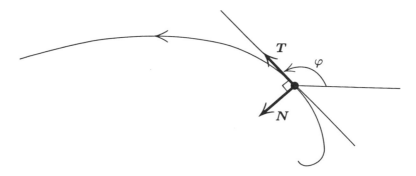

Figure V.1. Tangent and normal vectors

Let φ be the angle of inclination of T with respect to the positive x-axis. Then

(408) $T = (\cos\varphi, \sin\varphi)$ and $N = (-\sin\varphi, \cos\varphi)$.

It is clear that

(409) $$\frac{dT}{d\varphi} = N \quad \text{and} \quad \frac{dN}{d\varphi} = -T.$$

Use the chain rule to write

(410) $$\frac{dT}{ds} = k(s)N \quad \text{and} \quad \frac{dN}{ds} = -k(s)T,$$

where the function $k(s) = d\varphi/ds$ is called the *curvature* (function) of the curve. The curvature takes positive values if the curve is traversed in the direction of the tangent vector 'bends' toward the normal vector, and negative values in case the bending is away from the normal vector. Points where $k = 0$ are called *inflection points,* and the curve is a segment of a straight line exactly when $k(s) \equiv 0$.

The functions $ks, T(s), N(s)$ are called the *Frenet apparatus* of the curve. They satisfy the *Frenet equations* (410). Standard existence and uniqueness theorems for differential equations[8] allow us to conclude that knowing the continuous function $k(s)$, $a < s < b$, with specified $r(s_0)$ and $T(s_0)$ determines the curve uniquely for $a < s < b$. Stated another way,

---

[8] G.F. Simmons, *Differential Equations With Applications And Historical Notes, Second Edition,* McGraw-Hill, 1991, §70.

knowing $k = k(s)$, $a < s < b$, determines the curve up to congruence. For this reason, the specification $k = k(s)$ is called the *natural equation* of the curve.

A closed plane curve is called *convex* if any straight line cuts it in at most two points. Topological arguments imply that a convex plane curve is simple. A plane convex curve is called an *oval* if it is of class C^2 with respect to an arc length parameter and satisfies $k > 0$ at every one of its points. An oval has two orientations or directions of traversal. Following the tangent vector of the positive (and standard) orientation causes the oval to be traversed in the counterclockwise sense, and then the normal points into the interior component.

Assume that the coordinate origin lies in the interior component. Let α be the angle from the direction of the positive x-axis to the normal N, measured in the counterclockwise sense. Then $\alpha = \varphi + \frac{1}{2}\pi$, so that

$$(411) \qquad k = \frac{d\varphi}{ds} = \frac{d\alpha}{ds} > 0.$$

It follows from the Inverse Function Theorem that α, $0 \leq \alpha \leq 2\pi$, is also a regular parameter for the oval.

If (ξ, η) is a point on the oval with its normal at angle α, then the tangent line to the oval at this point has equation

$$(412) \qquad x \cos \alpha + y \sin \alpha + p(\alpha) = 0,$$

where

$$(413) \qquad p(\alpha) = -r(\alpha) \cdot N(\alpha)$$

is the *Minkowski support function* of the oval with respect to the specified origin. Clearly, $p(\alpha)$ depends upon the choice of the origin inside the oval, so it is not an invariant under congruence. Nevertheless, as we shall see, certain global properties of $p(\alpha)$ do not depend upon the choice of origin.

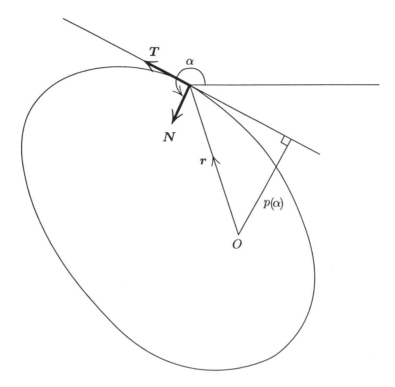

Figure V.2. Minkowski support function

We can recover the radius vector r from the support function. To simplify the typesetting, denote α-differentiation by a superdot. Because $\alpha = \varphi + \frac{1}{2}\pi$, (413) leads to

$$(414) \qquad \dot{p} = -\dot{r}{\cdot}N - r{\cdot}\dot{N} = -T{\cdot}N - r{\cdot}(-T) = r{\cdot}T.$$

Because T and N are orthogonal, we can resolve r along T and N to get

$$(415) \qquad r = (r{\cdot}T)T + (r{\cdot}N)N = \dot{p}T - pN.$$

We have

$$(416) \qquad \frac{1}{k}\frac{dr}{ds} = \frac{ds}{d\alpha}\frac{dr}{ds} = \frac{dr}{d\alpha} = (\ddot{p} + p)T.$$

Because $dr/ds = T$, we get

(417)
$$k = \frac{1}{\ddot{p} + p}.$$

EXERCISE V.7. If $r = (x, y)$, then

(418) $x = \dot{p}\sin\alpha - p\cos\alpha$ and $y = -\dot{p}\cos\alpha - p\sin\alpha.$

We get an $(\alpha, p(\alpha))$ description of the oval which should be compared to the $(r, p(r))$ description of a curve in Section 2.

All of the preceding has been 'local' because only derivatives of the various quantities are employed once the origin has been specified. Now we obtain a 'global' theorem, global in the sense that the hypotheses and conclusions can not be expressed only in terms of derivatives but depend upon the behavior of the curve over its whole extent.

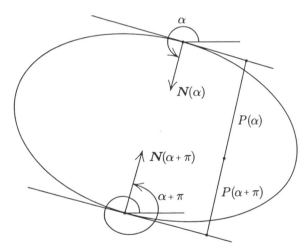

Figure V.3. Parallel tangents

Take an oval containing the origin within it, and let p be the corresponding Minkowski support function. Say that the oval has *constant width* if the distance between parallel tangent lines is constant—in particular, independent of their slope. If the normals to the two tangents point with the angles α and $\alpha + \pi$, then the associated width is $w(\alpha) = p(\alpha) + p(\alpha + \pi)$. (See Figure V.3.) The condition that the oval has constant width can be stated simply as $w(\alpha) \equiv w$, a constant. Clearly a circle has constant width with respect to its center, but it is not the only closed curve of constant width.

EXERCISE V.8. Prove that the following are closed, convex curves of constant width:

(a) The *Reuleaux triangles,* obtained from an equilateral triangle by replacing each side by an arc of a circle centered at the opposite vertex. These curves have interesting industrial application as the shapes of the rotors in the Wankel rotary engines, currently offered in certain Mazda automobiles.[9]

(b) The (smooth) parallel curves of the Reuleaux triangle, obtained as the envelopes of all circles of a fixed radius r whose centers range over the Reuleaux triangles. Such curves have an unexpected application to the construction of 'zig-zag' sewing machines.

We are now ready to prove our morsel from differential geometry in the large—*Barbier's Theorem,* published in 1880:

All ovals of constant width w have the same circumference πw.

With the machinery that we have built up in this section, proof of Barbier's Theorem is a short, direct calculation. Because $\ddot{p}$ is the derivative of $\dot{p}$ and $\dot{p}$ also has period 2π,

$$(419) \qquad \int_{\alpha=0}^{2\pi} ds = \int_0^{2\pi} \frac{ds}{d\alpha}\, d\alpha = \int_0^{2\pi} (\ddot{p} + p)\, d\alpha$$

$$= \int_0^{2\pi} p\, d\alpha = \int_0^{\pi} [p(\alpha) + p(\alpha + \pi)]\, d\alpha = \pi w.$$

6. Ovals described under a central attraction

Suppose that an oval can be described under the influence of a central attraction toward a point (the origin) inside the orbit. We know that the positive function $u(\vartheta) = 1/r(\vartheta)$ is related to the magnitude $P(r)$ of the attraction by the differential equation

$$(420) \qquad \frac{d^2 u}{d\vartheta^2} + u = \frac{P(1/u)}{h^2 u^2}.$$

The left side of this equation has the same form as the expression $\ddot{p} + p$ entering into the considerations of the previous section. This suggests introducing a new oval, the *companion to the orbit,* whose Minkowski

---

[9] R.F. Ansdale, *The Wankel RC Engine,* A.S. Barnes, 1969.

support function is $u(\alpha)$. According to Exercise V.7, we may describe the companion by the parametric equations

(421)
$$x = \dot{u}(\alpha)\sin\alpha - u(\alpha)\cos\alpha \quad \text{and} \quad y = -\dot{u}(\alpha)\cos\alpha - u(\alpha)\sin\alpha,$$

with period 2π in α.

EXERCISE V.9. Show that a circle is its own companion with respect to its center.

In order to transfer Barbier's Theorem of the previous section into a celestial context, we declare that a simple, closed orbit has *constant focal width* (with respect to an origin inside) if $u(\vartheta) + u(\vartheta + \pi)$ has a constant value on the orbit independent of the central angle ϑ. Now we can state our transferred theorem:

> If a simple, closed orbit is traversed under an attractive central force of magnitude $P(r)$ directed toward an interior point, and if the orbit has constant focal width w, then
>
> (422) $$\int_0^{2\pi} r(\vartheta)^2 P(r(\vartheta))\,d\vartheta = \pi w h^2,$$
>
> where h is the constant of angular momentum.

Because the transferred theorem is just a direct translation of Barbier's Theorem, we skip the proof and pass on to an example which shows that the hypotheses can be satisfied.

EXAMPLE. The ellipse (and oval) $r = l/(1 + e\cos\vartheta)$ is an orbit of period 2π for the attractive inverse square law with magnitude $P(r) = \mu/r^2$. Then $u(\vartheta) = (1 + e\cos\vartheta)/l$, so that $u(\vartheta) + u(\vartheta + \pi) = 2/l$, a constant. The ellipse therefore is an orbit of constant focal width for this central attraction. To check the conclusion of our theorem, note that $\pi w h^2 = 2\pi h^2/l$, while $\int_0^{2\pi} r^2 P(r)\,d\vartheta = \int_0^{2\pi} \mu\,d\vartheta = 2\pi\mu$. We know that $l = h^2/m u$, so the theorem is verified in this case.

EXERCISE V.10. Work out the detailed translation to prove the theorem.

EXERCISE V.11. Show that the companion to the elliptical orbit of the Example is a circle of radius $1/l$.

EXERCISE V.12. Find another example to illustrate the theorem of this section. (There are infinitely many geometrically different examples!)

EXERCISE V.13. We will call two points on an oval *antipodal* if the tangent vectors at the two points are antiparallel—they are each the negative of the other—and if the curvatures at the two points are equal. On a circle, an ellipse, or in general, any curve centrally symmetrical with respect to the origin, every pair of points antipodal in the usual sense by being at opposite ends of a diameter through the center is also antipodal in our new sense. According to a theorem proved by W. Blaschke and W. Süss,

> On every oval there are at least three antipodal pair of points.

There is a proof (with a minor glitch) in §16.3 of the book by Laugwitz cited earlier in this chapter.

(a) Find an oval with exactly three antipodal pairs of points. (*Hint:* Find an oval with the symmetries of the equilateral triangle and such that the curvature is monotonic on each of the six congruent subarcs.)

(b) Suppose that an oval is the orbit of a point under an attractive force of magnitude $P(r)$ toward an interior point. Compute the velocity vector of its companion and from it deduce that two points of the companion with parameters ϑ_1 and ϑ_2 have antiparallel tangents exactly when $\vartheta_2 = \vartheta_1 \pm \pi$.

Show further that the condition that companion has the same curvature at these two points is expressed by

$$(423) \qquad \ddot{u}(\vartheta_1) + u(\vartheta_1) = \ddot{u}(\vartheta_2) + u(\vartheta_2).$$

(c) Show that there are at least three pairs of *diametrically opposite* points (r_1, ϑ_1) and (r_2, ϑ_2) on the oval orbit for which

$$(424) \qquad r_1^2 P(r_1) = r_2^2 P(r_2).$$

This condition is of course satisfied identically for *any* two points on an elliptical orbit under Newton's Law of Universal Gravitation.

(d) What is a physical meaning for the quantity $r^2 P(r)$ that arose in this problem and in (422)?

Dynamical Properties of Rigid Bodies

1. From discrete to continuous distributions of mass

Fix a set of cartesian axes in space. Suppose a finite system of particles is distributed in space, the ith particle having mass m_i and position r_i. The *center of mass* of the system is defined to be the point $\bar{r}$, where

$$(425) \qquad \bar{r} = \frac{\sum m_i r_i}{\sum m_i}$$

and the summation is over all of the particles. Being defined vectorially, the center of mass is independent of the orientation of the axes in space; it also is easy to see that it is independent of the choice of origin.

The *center of gravity* $\bar{r}_g$ is defined to be that point such that the sum of the moments of the forces $m_i g$ acting at r_i is equal to the moment of the combined mass $\sum m_i$ acting at r_g: that is,

$$(426) \qquad r_g \times \left(\sum m_i \right) g = \sum \left(r_i \times (m_i g) \right).$$

This equation has the general solution

$$(427) \qquad r_g = \frac{\sum m_i r_i}{\sum m_i} + kg = \bar{r} + kg,$$

where k is an arbitrary scalar. The specification of r_g by the equation $r_g = \bar{r} + kg$ will be independent of the direction of g only if $k = 0$. Therefore, we specify that $r_g = \bar{r}$.

To extend the definitions to continuous distributions of mass, we 'smear' the particle mass m_i over a small bit of space Δv_i and we suppose $\bar{\rho}_i$ to be the average density in that region. Calling the mass now Δm_i, we suppose that $\Delta m_i = \bar{\rho}_i \Delta v_i$. Then

$$(428) \qquad\qquad \bar{r} = \frac{\sum (\Delta m_i) \bar{r}_i}{\sum \Delta m_i}$$

(where $\bar{r}_i$ is a point within Δv_i) has some claim to be an approximate center of mass. There is, however, some ambiguity in (428). We remove the ambiguity by noticing that (428) contains the quotient of Riemann sums for two integrals. We *define*, therefore, the *center of mass* by

$$(429) \qquad\qquad \bar{r} = \frac{\int r \, dm}{\int dm},$$

the integrals extending over the whole body. These are actually Riemann-Stieltjes integrals. If we put $dm = \rho \, dv$, ρ being the density, then

$$(430) \qquad\qquad \bar{r} = \frac{\int r \rho \, dv}{\int \rho \, dv}.$$

It is easy to see that if the solid body is symmetrical with respect about some element (plane, line, point) both geometrically and with respect to density or mass, then the center of mass lies upon that element.

EXERCISE VI.1. Prove the last assertion.

2. Moments of inertia

It will be convenient to use index notation in this section. Let the rectangular coordinates be (x_1, x_2, x_3). The results will apply to discrete or to continuous distributions of mass, the signs $\sum$ or $\int$ to be used as appropriate. For example, $\sum m_i x_i^2$ and $\int x^2 \, dm$ will correspond.

Define six quantities I_{ij} by

(431)
$$I_{11} = \int (x_2^2 + x_3^2)\, dm,$$
$$I_{22} = \int (x_3^2 + x_1^2)\, dm,$$
$$I_{33} = \int (x_1^2 + x_2^2)\, dm,$$
$$I_{23} = -\int x_2 x_3 \, dm,$$
$$I_{31} = -\int x_3 x_1 \, dm,$$
$$I_{12} = -\int x_1 x_2 \, dm.$$

We sometimes write

(432)
$$I_{11} = A, \quad I_{22} = B, \quad I_{33} = C,$$
$$I_{23} = -F, \quad I_{31} = -G, \quad I_{12} = -H.$$

Then A, B, C are the *moments of inertia* and F, G, H are the *products of inertia* of the system of mass.

Let the system have angular velocity $\boldsymbol{\omega}$, so that the mass element dm at $\boldsymbol{r}$ has velocity $\boldsymbol{\omega} \times \boldsymbol{r}$ and angular momentum around the origin $\boldsymbol{r} \times (\boldsymbol{\omega} \times \boldsymbol{r})\, dm$. The total angular momentum of the system is

(433)
$$\boldsymbol{h} = \int \boldsymbol{r} \times (\boldsymbol{\omega} \times \boldsymbol{r})\, dm$$
$$= \int [r^2 \boldsymbol{\omega} - (\boldsymbol{r} \cdot \boldsymbol{\omega})\boldsymbol{r}]\, dm.$$

The vector $\boldsymbol{h}$ has components (h_1, h_2, h_3); resolving the last equation, we find that

(434)
$$h_i = \sum_j \omega_j I_{ji}.$$

Introduce vectors and matrices

(435)
$$\boldsymbol{\omega} = \begin{bmatrix} \omega_1 \\ \omega_2 \\ \omega_3 \end{bmatrix}, \quad \boldsymbol{h} = \begin{bmatrix} h_1 \\ h_2 \\ h_3 \end{bmatrix}, \quad I = \begin{bmatrix} I_{11} & I_{12} & I_{13} \\ I_{21} & I_{22} & I_{23} \\ I_{31} & I_{32} & I_{33} \end{bmatrix},$$

where the matrix I is symmetric: $I_{\beta\alpha} = I_{\alpha\beta}$. Then

(436) $$h^T = \omega^T I \quad \text{or} \quad h = I\omega$$

in terms of matrix multiplication.

The matrix I is a symmetric tensor of rank two: in the present case, the important property is that the quantity

(437) $$h^T \omega = \omega^T I \omega$$

is a symmetric quadratic form in ω whose value is independent of the coordinate system in which ω and I are expressed. (This is straightforward to check.) It follows that there is a preferred system of coordinate axes in which the quadratic form

(438) $$Ax^2 + By^2 + Cy^2 - 2Fxy - 2Gzx - 2Hyz$$

—which defines the momental ellipsoid with respect to the origin—takes on diagonal form. These are the *principal axes*.

3. Particular moments of inertia

For the homogeneous sphere, symmetry implies that $A = B = C$. If ρ is the (constant) density, then

(439) $$A + B + C = 2 \int (x^2 + y^2 + z^2)\, dm = 2\rho \int r^2\, dv.$$

Use spherical coordinates, so that $dv = r^2\, d\omega$, where $d\omega$ is the element of solid angle. Then

(440) $$A + B + C = 2\rho \cdot 4\pi \int_0^a r^4\, dr = \tfrac{8}{5}\pi\rho a^5,$$

where a is the radius of the sphere. Hence,

(441) $$A = B = C = \tfrac{8}{15}\pi\rho a^5 = \tfrac{2}{5}Ma^2,$$

where M is the mass of the sphere.

The axes of a homogeneous ellipsoid are its principal axes of inertia. Then

(442) $$A = \int (y^2 + z^2)\, dm = \rho \int (y^2 + z^2)\, dv,$$

integrating over the volume contained in the ellipsoid

(443)
$$\frac{x^2}{a^2} + \frac{y^2}{b^2} + \frac{z^2}{c^2} = 1.$$

Change variables by $\xi = x/a$, $\eta = y/b$, and $\zeta = z/c$. The integration now extends over the ball $\xi^2 + \eta^2 + \zeta^2 = 1$. It is an easy calculation to arrive at

(444)
$$A = \frac{1}{5}M(b^2 + c^2)$$

and—by notational symmetry!—

$$B = \frac{1}{5}M(c^2 + a^2),$$
$$C = \frac{1}{5}M(a^2 + b^2),$$

where M is the mass of the ellipsoid,

EXERCISE VI.2. (a) Prove that the moment of inertia of a body with center of gravity G about any axis is equal to the moment of inertia about a parallel axis through G plus Mp^2, where M is the mass of the body and p is the distance from G to that axis.

(b) Find the moment of inertia of an elliptic cylinder about a generator through the end of a major axis and through the end of a minor axis, as well as an axis through a focus.

(c) Find principal axes and moments of inertia of a homogeneous sphere rotating around a point on its surface.

EXERCISE VI.3. A body has angular velocity ω about an axis passing through the origin. Show that its kinetic energy is $\frac{1}{2}\omega^T I\omega$. If the center of mass is not at the origin, decompose the kinetic energy into the sum of two terms, one representing rotation around the parallel axis through the center of mass, the other giving the energy of rotation of the center of mass around the axis of rotation.

EXERCISE VI.4. The moments of inertia about the principal axes of simple bodies are given by a scheme known as *Routh's rule,* according to which they are Ms^2/n, where s^2 is the sum of the squares of the semiaxes of the body perpendicular to the given axis, and n has the value $3, 4$, or 5 according to whether the body is rectangular, elliptical, or ellipsoidal.

4. Euler's equations of motion

As the body moves, the expression of I in terms of fixed (inertial) axes will vary, but it will remain unchanged with respect to axes *fixed in the body*. From (31) we know that the rate of change of angular momentum with respect to time is torque. Let the moment of the external forces on the body be Γ, and suppose that ω is the instantaneous angular velocity. Then, from (10),

$$
(445) \qquad \begin{aligned}
\Gamma &= \frac{d}{dt}(I\omega) \\
&= \frac{\partial}{\partial t}(I\omega) + \omega \times (I\omega) \\
&= I\dot{\omega} + \omega \times (I\omega).
\end{aligned}
$$

If the axes fixed in the body are taken to be principal axes, then

$$
(446) \qquad I = \begin{bmatrix} A & 0 & 0 \\ 0 & B & 0 \\ 0 & 0 & C \end{bmatrix},
$$

and the equations of motion are written simply in component form as

$$
\begin{aligned}
(447) \qquad \Gamma_1 &= A\dot{\omega}_1 - (B - C)\omega_2\omega_3, \\
(448) \qquad \Gamma_2 &= B\dot{\omega}_2 - (C - A)\omega_3\omega_1, \\
(449) \qquad \Gamma_3 &= C\dot{\omega}_3 - (A - B)\omega_1\omega_2.
\end{aligned}
$$

These are *Euler's equations of motion*. They have an attractive, cyclical form.

5. Euler free motion of the Earth

Newton proved[1] that the Earth has the form of an oblate spheroid with axes a, a, c, where $a > c$. The best current value for the flattening $\epsilon = (a-c)/a$ is about $1/297$. Nevertheless, there remained those who doubted Newton's proof. Starting in 1683 and continuing into the 1730s, four generations of the Cassini family, along with other notables, carried out surveys along meridians in France which led them to the conclusion that

---

[1] *Principia Mathematica*, Book III, Proposition XVIII.

the Earth must be a prolate spheroid. The resulting controversy was settled in favor of the oblate by the French Academy of Sciences, which in the late 1730s sent expeditions to Lapland (Maupertuis, Clairaut, Celsius, et al.) and to Peru (Godin, La Condamine, Bouguer, et al.) to measure the length of a degree of longitude.[2]

The undoubted nonsphericity of the Earth has many dynamical consequences. We can illustrate the use of the Euler equations by predicting the so-called *Euler free motion* of the Earth.

Consider the rotation of the Earth, ignoring all outside forces acting on it. Suppose the Earth to be a rigid spheroid with principal moments of inertia A, B, C. Then $A = B < C$.

We specialize the Euler equations (447) by setting $B = A$ and $\Gamma = 0$. There are no outside forces, so there is no torque. Then

$$
\text{(450)} \qquad A\dot{\omega}_1 = (A - C)\omega_2\omega_3,
$$

$$
\text{(451)} \qquad A\dot{\omega}_2 = (C - A)\omega_3\omega_1,
$$

$$
\text{(452)} \qquad C\dot{\omega}_3 = 0.
$$

From the third of these equations, we have $\omega_3 = $ constant, so that the (idealized) Earth spins about its polar axis of symmetry at a constant rate.

Let

$$
\text{(453)} \qquad n = \frac{C - A}{A}\omega_3.
$$

Then the first two equations become

$$
\text{(454)} \qquad \dot{\omega}_1 + n\omega_2 = 0 \quad \text{and} \quad \dot{\omega}_2 - n\omega_1 = 0.
$$

It is easy to solve this system and find that

$$
\text{(455)} \qquad \omega_1 = \alpha\cos(nt + \beta) \quad \text{and} \quad \omega_2 = \alpha\sin(nt + \beta).
$$

(The amplitudes and the phases agree.) If we measure time in days, then $\omega_3 = 2\pi$, and so the period of ω_1 and ω_2 is

$$
\text{(456)} \qquad \frac{2\pi}{n} = \frac{A}{C - A}.
$$

---

[2] J.R. Smith, *From Plane To Spheroid*, Landmark Enterprises, 1986. The hardships that the members of these expeditions endured make their stories into sagas.

The value of this quotient, the *Eulerian free period,* is known to be about 303 sidereal days.

There is a kinematical consequence. Suppose that unit vectors i, j, k point from the center of mass along the principal axes, with k toward the 'north pole.' Then the angular velocity vector is $\omega = \omega_1 i + \omega_2 j + \omega_3 k$. Because we are assuming that the Earth is a rigid body, the law of conservation of angular momentum requires that ω be fixed in space (that is, with respect to an inertial coordinate frame).

Define axial angular velocity vectors $\omega_1 = \omega_1 i$, $\omega_2 = \omega_2 j$, and $\omega_3 = \omega_3 k$. Then $\omega_3 \cdot \omega = \omega_3 = $ constant, so that the angle from ω_3 to ω remains constant in time. The vector $v = \omega_3 \times \omega$ is perpendicular to k. In fact,

$$(457) \quad v = \begin{vmatrix} i & j & k \\ 0 & 0 & \omega_3 \\ \omega_1 & \omega_2 & \omega_3 \end{vmatrix} = \omega_3 \alpha \{ -\sin(nt + \beta)i + \cos(nt + \beta)j \}.$$

It is clear from this representation that v turns steadily around the k axis with period $2\pi/n$, the Eulerian free period.

There is an important interpretation of this periodic motion. With respect to the axes i, j, k fixed in the Earth, the vector ω rotates around the polar k-axis with angular velocity n, sweeping over the surface of a cone. But ω remains fixed with respect to the 'fixed' stars. The practical consequence is that the apparent astronomical latitude of a point fixed on the surface of the Earth should oscillate north and south with period $2\pi/n \approx 10$ months.

The apparent latitude is determined empirically by collating astronomical observations from a world-wide network of sites. It is found that a single periodic term does not suffice for representing the empirical variations in latitude. Two periodic terms suffice to a high degree of satisfaction. The first has amplitude $0''.09$ and period about one year. The second—the *Chandler wobble*—has amplitude $0''.18$ and period about 14 months. There are, in fact, torques acting on the Earth. Among the causes of external forces and their torques are changes in the atmosphere from seasonal and other activities; tidal forces from water on the surface of the Earth; and forces within the Earth from shifting of crustal plates and viscous sloshing of a molten core. There are also torques caused by gravitational action of the Moon and the Sun on the 'bulges' around the Earth's equator. These torques induce the *lunar-solar precession*, also called the precession of the equinoxes, which we will discuss in Section VII.3.

In any event, our analysis produces only the period of the Eulerian free oscillation; it may be that the amplitude of the oscillation—which can not be determined from the differential equations (450)—is just too small to be observed.

6. Feynman's wobbling plate

The physicist Richard Feynman began his ascent up the academic ladder as an assistant professor at Cornell University. He preferred to lunch at the cafeteria rather than at the faculty club because he liked to look at the pretty girls. As he told his biographer Jagdish Mehra,[3] one day he noticed a student playing with a plate, tossing it into the air and catching it. The plate had a medallion with an effigy of Ezra Cornell, the founder of the university, printed on it.

> As the plate went up into the air, I saw it wobble, and I noticed that the medallion on the plate was going around. It was pretty obvious to me that the medallion went around faster than the wobbling. I had nothing to do, so I started to figure out the motion of the rotating plate. I discovered that when the angle [of the wobble with the horizontal] is [very small], the medallion rotates twice as fast as the wobble rate—two to one. It's a cute relationship; it came out of a complicated equation. ... I wanted to understand this motion from Newton's laws alone. I wanted to see the forces, ... I wanted to see how [Newton's] laws of motion applied to the disk.

We can recover Feynman's conclusions readily from what we have already worked out. In fact, the discussion of the Euler free motion in

---

[3] Jagdish Mehra, *The Beat of a Different Drum: the Life and Science of Richard Feynman*, Oxford University Press, 1994, p. 173 and pp. 180–81.

Section 5) includes all of the apparatus that we need here. The derivation of Euler free motion, nominally about the Earth, used only certain symmetries of the Earth's figure together with the values of the principal moments of inertia. It applies equally well to a disk. Feynman's wobbling plate is simply the Euler free motion for the *flat Earth* of the early Greek philosophers.[4]

The disc has principal moments of inertia $A = B$ and C. It is easy to work out directly or it follows from Routh's rule (Exercise VI.4)) that $C = 2A$. Then (453), which relates the angular velocity ω_3 of the plate around its 'polar' axis and its overall angular velocity n, reduces to $n = \omega_3$. But the conserved angular velocity vector with respect to a Newtonian inertial frame is the sum of the corresponding angular velocity vectors of the two separate rotations. If the wobble is 'small,' then the two separate angular velocity vectors will be nearly parallel. Because angular velocity vectors combine by vector addition (p. 3), the length of the resultant conserved angular velocity vector will be very close to $n + \omega_3 = 2\omega_3$. This is what Feynman saw: The period of the spin is half the period of the wobble.[5]

EXERCISE VI.5. Obtain a nonfragile, discoidal object and decorate it with a suitable effigy. Toss the object spinning into the air and replicate Feynman's observations.

EXERCISE VI.6. A *baton* is a circular-cylindrical rod of uniform density whose diameter is negligible compared to its length. Examine the free motion of a baton tossed into the air, including the question of whether it can move with a wobble analogous to the wobble of Feynman's plate. There are at least two cases, depending upon whether the baton is or is not rotating around its long axis.

Perform some experiments in baton twirling to verify your predictions. (An ordinary wood pencil might make a satisfactory experimental baton.)

---

[4] 'Ancient' is appropriate here. Contrary to current popular opinion, the 'roundness' of the Earth was accepted long ago by the Greek philosophers. What distinguishes Christopher Columbus is not his championing of the round Earth over the flat Earth, but his insistence that the Earth's circumference is much smaller than was commonly accepted in his time. See J.B. Russell, *Inventing the Flat Earth: Columbus and Modern Historians,* Praeger, 1991.

[5] Jagdish Mehra, *The Beat of a Different Drum: The Life and Science of Richard Feynman,* Oxford University Press, 1994, p. 179, note 32. See also the note by Benjamin Fong Chao, *Physics Today,* February 1989, p. 15, mentioned by Mehra.

7. The gyrocompass

In ancient times, long ocean voyages usually were carried out as a succession of daytime stages within sight of land, with nightly layovers on shore. This kind of sailing required experienced pilots and carried the constant peril that the ship would be caught in unexpected currents or that it would be dashed upon the rocks during a sudden storm. Guidance by the Sun during the day and the stars at night was undependable, given the possibility of clouds obscuring the sky for long periods.

The discovery of the magnetic compass opened up the possibility of deep water navigation. However, the magnetic compass was soon seen to be unreliable for guidance at high latitudes. Moreover, it was subject to a natural error and unavoidable error because of the phenomenon of *magnetic declension,* which was mapped over the Atlantic Ocean in 1700 by Captain Edmond Halley, Royal Navy, during his voyage of scientific discovery in His Majesty's pink, the *Paramore.*[6] Some of the problems associated with the magnetic compass were solved by the introduction of the gyroscopic compass, which orients itself with respect to an inertial frame in space.

We give a simplified explanation of the operation of the gyrocompass, which consists of a rapidly spinning disk D (see Figure VI.1), suitably suspended so that its center of mass remains fixed as the disk and axle move.

The disk rotates on an axle kept horizontal by a suitable mounting. Take the k-axis along the rotation axis of the disk, and suppose the angular speed of rotation to be a large number s. The disk is cylindrically symmetrical around the rotation axis, so that it is not necessary for the i, j vectors to rotate with the disk in order for the moments of inertia A, B ($A = B$) around those axes to remain constant. Take j vertical and i in the horizontal plane perpendicular to k, so that the frame i, j, k has right-handed orientation.

---

[6]Halley's observations were summarized in an isogonic chart, showing curves on which the magnetic declination is constant. He invented the isoline chart.

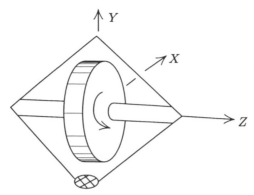

Figure VI.1. The gyrocompass (in idealized suspension)

Represent by Ω the angular speed of the Earth's rotation, by λ the latitude, and by φ the angle measured positive toward the west that $\boldsymbol{k}$ makes with the meridian. The vertical component of the angular velocity of the Earth is $\Omega \sin \lambda$, and the northward horizontal component is $\Omega \cos \lambda$. We can resolve the angular velocity of the frame $\boldsymbol{i}, \boldsymbol{j}, \boldsymbol{k}$ with respect to itself as

$$(458) \qquad \begin{aligned} \omega_1 &= -\Omega \cos \lambda \sin \varphi, \\ \omega_2 &= \Omega \sin \lambda + \dot{\varphi}, \\ \omega_3 &= \Omega \cos \lambda \cos \varphi. \end{aligned}$$

The total angular velocity around the $\boldsymbol{k}$ axis will be the sum of ω_3, from the motion of the $\boldsymbol{i}, \boldsymbol{j}, \boldsymbol{k}$ frame, and s, from the axial rotation of the gyroscope. We may therefore use the Euler equations (447) after we replace every occurrence of ω_3 by $\omega_3 + s$.

Consider motion around the vertical direction $\boldsymbol{j}$. There is no torque acting around this axis, so $\Gamma_2 = 0$. Substitute the expressions (458) for the angular velocity components into the second Euler equation from (447) with the modifications just described,

$$(459) \qquad A\dot{\omega}_2 - (C - A)\omega_3\omega_1 - Cs\omega_1 = 0,$$

and simplify to get

$$(460) \quad A\ddot{\varphi} + (C - A)\Omega^2 \cos^2 \lambda \cos \varphi \sin \varphi + Cs\Omega \cos \lambda \sin \varphi = 0.$$

The gyroscope is driven at a high angular speed while Ω is 2π per day, so we can neglect Ω in comparison with s. Therefore, we drop the Ω^2-term, divide by A, and consider the differential equation

(461) $$\ddot{\varphi} + (C/A)s\Omega\cos\lambda\sin\varphi = 0.$$

Compare this differential equation with the differential equation (55) for the oscillation of a simple pendulum. The gyrocompass will oscillate around the vertical direction j with a period P for *small* amplitudes essentially given by

(462) $$P = 2\pi\sqrt{\frac{A}{Cs\Omega\cos\lambda}}.$$

Because $\varphi = 0$ is the direction of the meridian, the gyrocompass will, if there is some damping, come to rest pointing toward true geographical north.[7]

Note that the period of the undamped gyrocompass has essentially infinite duration near the poles of the Earth, but we have ignored the practical aspects of suspension and damping.

EXERCISE VI.7. Obtain a gyroscope, rig a suspension that holds the axle horizontal yet allows the axis to swing in the horizontal plane, and try to observe the oscillations predicted by (461). Note that, as you experiment, you can not expect $\dot{s} = 0$ for long.

8. Euler angles

We want to coordinatize the rotations of three-space; these are the maps of R^3 into itself that leave the origin fixed and preserve both Euclidean distances and orientation. This will be done for its own interest and also for application in the next section.

First of all, ignoring orientation, the maps that leave the origin fixed and preserve Euclidean distance are the same as those that preserve inner products: they are maps $T : R^3 \to R^3$ such that $T(0) = 0$ and for which $T(u) \cdot T(v) = u \cdot v$ for all vectors u, v. One can show that such a map T is linear. Furthermore, it can be expressed with respect to the standard basis for R^3 as multiplication by a matrix, $T(v) = Ov$, where $OO^{-1} = I$, the

---

[7]Of course, true geographical north is not fixed in time, but wanders over the years due to as yet not fully understood influences within a rectangle a few hundred meters across.

identity matrix. Such transformations and their associated matrices are called *orthogonal*. In particular, the identity transformation is orthogonal and it is represented by the identity matrix I.

Call an orthogonal transformation T a *rotation* if T can be joined to the identity by a continuous path of orthogonal matrices.[8]

EXERCISE VI.8. (a) Show that the transformation with matrix $-I$ is not a rotation of R^3.

(b) Show that composition, inverse, and transpose on rotations result in rotations.

(c) The rotations of R^3 form a group, called $R(3)$.

We will now prove that every rotation turns R^3 rigidly through a certain angle around an axis. This can be done easily—once the paraphernalia of matrix theory has been set up—by showing that every skew symmetric 3×3 matrix has an eigenvalue ± 1. However, there is a simple, geometric proof due to Euler.

In the generic case, let P and Q be distinct, nonantipodal points on the unit sphere S^2 in R^3. Suppose that $T(P) = P'$ and $T(Q) = Q'$, the points P', Q' being again on S^2 and not both P, Q, respectively. Then the great circle G determined by P, Q must be mapped onto the distinct great circle G' through P', Q'. Two distinct great circles intersect in antipodal 'poles' N, S. Then the rotation leaves N, S fixed and rotates the plane of G through a certain angle (between $-\pi$ and π) around the line (axis) through N, S. The nongeneric cases are treated similarly and more simply.

EXERCISE VI.9. Show that the set of rotations of R^2 into itself is in natural, one-to-one correspondence with the points of the unit circle.

Can we 'visualize' the set of rotations of R^3? Plot a rotation in 'polar coordinates' in R^3: If its rotation angle is α (between $-\pi$ and π) around the polar semi-axis through N, represent the rotation by a point at distance α from the origin along the same polar semi-axis. Of course, we have no way of distinguishing the two points N, S, and the same rotation can be represented by the point along the polar semi-axis through S, with the rotation being in the opposite sense. We get a unique representative of the physical rotation when the angle of rotation is $\neq \pm\pi$, but the

---

[8]Say that $\lim_{t \to t_0} T(t) = T^*$ if $\lim_{t \to t_0} T(t)(v) = T^*(v)$ for every vector v.

rotation through π around an axis is physically the same as rotation about its antipodal axis through angle $-\pi$. Therefore, we conclude that the set of rotations in R^3 is represented by the closed sphere of radius 2π on the surface of which antipodal points are identified ('glued' together). The resulting object is P^3, the real projective three-space, a compact topological space that is much more complicated than the circle which represents rotations of R^2.

The coordinatization of a circle by a central angle is non-holonomic: a tracing point that leaves a starting point and goes around the circle in a clockwise direction returns to the starting point with its angle the starting angle plus 2π. The situation for rotations of R^3 is much more complicated, because P^3 is three-dimensional and three coordinates are required to locate points in it. There are many ambiguities in any coordinate prescription of P^3, and topologists have proved that there is no way to avoid these ambiguities or even to reduce them to even the 'simple' ambiguity of the angular coordinatization of the circle. This has important practical consequences for gyroscopic devices such as inertial navigation systems in airplanes.[9]

We will describe a coordinatization of rotations in R^3 that is called *Euler angles*. In fact, there are many possible definitions of systems of Euler angles, but they all have similar structures.

We describe Euler angles with reference to Figure VI.2.

Let i_0, j_0, k_0 be a fixed system of orthogonal axes, which a rotation takes into orthogonal axes i, j, k. We can realize the rotation in three successive steps:

(1) Rotate the frame i_0, j_0, k_0 through an angle ψ around the k_0-axis, resulting in a frame i_1, j_1, k_1, where $k_1 = k_0$. We may take, say, $0 \leq \psi \leq 2\pi$.

(2) Rotate the frame i_1, j_1, k_1 through an angle ϑ around the i_1-axis, resulting in a frame i_2, j_2, k_2, where $i_2 = i_1$. We may take, say, $0 \leq \vartheta \leq \pi$.

(3) Rotate the frame i_2, j_2, k_2 through an angle φ around the k_2-axis,

---

[9] Kenneth R. Britting, *Inertial Navigation Systems Analysis*, Wiley, 1971.

resulting in the frame i_3, j_3, k_3, where $i_3 = i, j_3 = j$, and $k_3 = k$. We may take, say, $0 \leq \varphi \leq 2\pi$.

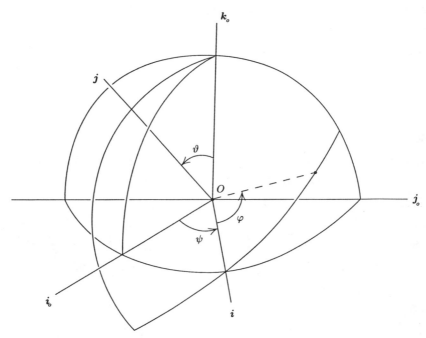

Figure VI.2. Euler angles

The endpoints of the domain intervals give many sets of coordinates to the same point and, of course, the angles may be restricted to other intervals of the same length. The quantity $d\psi/dt$ is the angular velocity with which the frame i, j, k is turning around the k-axis. There are important interpretations as angular velocities also for $d\vartheta/dt$ and $d\varphi/dt$.

There are schemes for 'cleaning up' the ambiguities and non-symmetries of the Euler angles.[10] From the topological point of view, the rotation group is double-covered by S^3, the three-dimensional unit sphere in R^4. The sphere S^3 can be viewed as the set of unit *quaternions* and as such has a natural multiplication on it. The algebra of rotations in R^3 can be 'lifted' to the multiplicative algebra of the unit quaternions, which has a great deal more symmetry than does $R(3)$. The development of

---

[10]The following discussion requires more advanced mathematics than is assumed for the rest of this book.

quaternion algebra, discovered by W.R. Hamilton in 1843, resulted in a thriving research industry, but the 'doubleness' of the covering of $R(3)$ by S^3 complicated the application of quaternion algebra to 'real' physics problems. In 1881 and 1884, J.W. Gibbs (privately) published his vector analysis, which vanquished the more complicated quaternion methods, although not without a drawn-out struggle.[11] Nevertheless, quaternion methods remain of some use, particularly when the goal is a symmetric presentation of an intrinsically symmetric process. For example, they are effective in dealing with the kinematics of a planet in orbit around the Sun.[12]

---

[11]Michael J. Crowe, *A History of Vector Analysis,* University of Notre Dame Press, 1967. Available with updated references from Dover Publications, (1985).

[12]Eduard L. Stiefel and G. Scheifele, *Linear and Regular Celestial Mechanics,* Springer-Verlag, 1971.

VII

Gravitational Properties of Solids

1. The gravitational potential of a sphere

The potential energy at O of a particle of mass m at P is $-Gm/r$. The potential produced at O by a solid body with density function ρ is

$$(463) \qquad V = -G \int \frac{\rho \, dv}{r}.$$

In both cases, r is the distance from the current point to O. The integral formula comes from approximating the solid body by a system of particles, whose gravitational potential energy is a Riemann sum for the integral. We suppose that the density is everywhere finite. Then the integrand is finite everywhere if O is outside the body. If O is inside or on the surface, then $1/r$ becomes infinite somewhere. The integral is then improper and is defined to be the limit as the radius goes to zero of the integral obtained by excluding a small ball centered at O. Introduce polar coordinates centered at O. The element of volume is $r^2 \, dr \, d\omega$. Then

$$(464) \qquad V = -G \int \rho r \, dr \, d\omega,$$

and the integral converges.

Consider the attraction produced at an internal point O by a homogeneous shell bounded by concentric spheres. Suppose a cone with vertex O cuts the shell in two frustums $PQQ'P'$ and $RSS'R'$. (See Figure VII.1.)

Let ρ be the density. Then the mass of a thin slice of cone of thickness dr is $\rho r^2\, dr\, d\omega$, and its attraction at O is essentially $G\rho\, dr\, d\omega$. Therefore, the attractions of the frustums are essentially $G\rho\, d\omega\, PQ$ and $G\rho\, d\omega\, RS$. But the bounding spheres are concentric, so they make equal intercepts on each chord. Therefore, $PQ = RS$ and the frustums cause equal and opposite attractions at O. Integrating over the entire solid angle at O, we find the resultant attraction to be zero.[1]

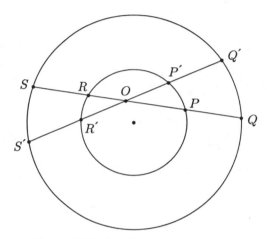

Figure VII.1. Concentric spheres

Call two ellipsoids *similar* if their axes are parallel, concentric, and of constant ratio. A uniform shell bounded by two similar ellipsoids is called a *homoid.* Such a figure can be considered to be an oblique projection of the figure bounded by two spheres. Hence, the attraction of a homoid upon an internal point is again zero.

Now consider the potential of a homogeneous, thin, spherical shell at an external point O. Let the shell have center C, radius a, and thickness da. If P is a point in the shell (Figure VII.2), let $\vartheta = \angle OCP$. Consider a thin ring perpendicular to CP and lying between ϑ and $\vartheta + d\vartheta$. Its radius is $a \sin \vartheta$ and its mass is $(\rho)(2\pi a \sin \vartheta)(a\, d\vartheta\, da)$, where ρ is the density of the shell. Any element of the ring is at distance $l = \sqrt{r^2 + a^2 - 2ar \cos \vartheta}$

---

[1] This argument, due to Newton, appears to neglect possibly significant odd-shaped pieces. They are of 'higher order' and their neglect can be justified easily. Of course, the result follows by brute-force integration using the formula (463).

from O, so that the potential of the ring at O is

(465)
$$-G\rho \cdot 2\pi a^2\, da \, \frac{\sin\vartheta\, d\vartheta}{\sqrt{r^2 + a^2 - 2ar\cos\vartheta}}$$

and the total potential at O caused by the entire shell is

(466)
$$dV = -G\rho \cdot 2\pi a^2\, da \int_0^\pi \frac{\sin\vartheta\, d\vartheta}{\sqrt{r^2 + a^2 - 2ar\cos\vartheta}}.$$

But $\rho \cdot 4\pi a^2\, da = dm$, the mass of the shell. Therefore,

(467)
$$dV = -\tfrac{1}{2}G\, dm \left[\frac{1}{ra}\sqrt{r^2 + a^2 - 2ar\cos\vartheta} \right]_0^\pi$$
$$= -\frac{G\, dm}{r},$$

because $r > a$. This means that the shell produces the same gravitational effect at O as it would if all its mass were concentrated at its center.

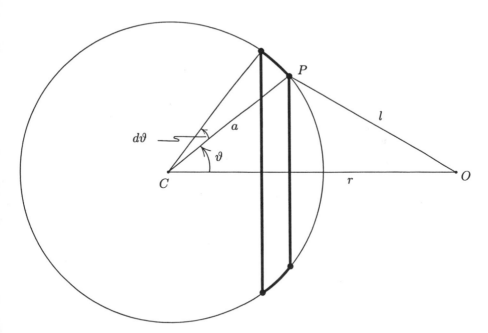

Figure VII.2. Ring element of a spherical ball

To obtain the potential at an external point O of a solid sphere of constant density ρ and center C, consider the sphere to be composed of concentric shells. The shell of radius α and thickness $d\alpha$ has mass

$dm = 4\pi\rho\alpha^2\, d\alpha$ and potential $dV = -(G\rho/r) \cdot 4\pi\alpha^2\, d\alpha$. Therefore, the potential produced at O by the sphere is

$$(468) \qquad\qquad V = -\int_0^a \frac{G\rho}{r} 4\pi\alpha^2\, d\alpha = -\frac{GM}{r},$$

with M the total mass of the sphere. The effect at O is as if all of the mass were concentrated at the center. (This is a partial answer to the question raised in the footnote on page 44.) The same effect is produced at an external point by an inhomogeneous sphere, provided that the density is a function of the distance from the center of the sphere.

EXERCISE VII.1. Compare the magnitudes of the gravitational attractions at the Earth's surface of the Earth itself, its moon, the Sun, the remaining eight planets, and of a 160-lb obstetrician who is at arm's length on a baby being delivered. Assume that all of the heavenly bodies are in a line on the same side of the Sun. Give the results as order of magnitude estimates with the values normalized so that the attraction of the Earth itself is 1.

EXERCISE VII.2. Show that the initial rate of increase of g in descending a mine shaft will be equal to g/a if the density ρ of the Earth were uniform, a being the radius of the Earth. But if the Earth had a spherical nucleus of different density and radius b, the density of this nucleus must be

$$(469) \qquad\qquad \rho\left\{1 + \frac{a^3(1-\lambda)}{b^3(2+\lambda)}\right\},$$

where $\lambda g_{new}/a$ is the initial rate of decrease of g_{new}, the new acceleration of gravity, in descending the shaft.

EXERCISE VII.3. Express the gravitational potential at a point within a homogeneous sphere and at a distance r from its center as a function of r and the total mass M. (Note that the potential must be continuous across the surface of the sphere.)

EXERCISE VII.4. Find the law of density $\rho = \rho(r)$ in a spherically symmetric solid when the attracting force at an internal point of the sphere at distance r from the center has the magnitude $F(r)$. Under what $\rho(r)$ can the magnitude $F(r)$ be constant?

EXERCISE VII.5. A uniform solid sphere of mass M is placed near an infinite plate whose surface density is uniform and equal to σ. Prove that the sphere attracts the plate with a force of intensity $2\pi GM\sigma$.

2. Potential of a distant body; MacCullagh's formula

The list of figures for which exact expressions have been obtained for the potentials is appallingly small. Even homogeneous ellipsoids call for tedious calculations. If, however, only the far field—the potential at points greatly distant from the body—is needed, there is a simple approximation for the potential due to MacCullagh. The principal terms are, in a sense, independent of local variations in the shape of the body and of local perturbations in the distribution of mass.

Let G be the center of gravity of the body, O a point at distance R from G, where R is great in comparison to the dimensions of the body. (See Figure VII.3.) Suppose that a particle of mass dm is at P, at a distance r from G, such that ϑ is the angle $\angle OGP$. Then the potential at O is

$$
(470)\quad V = -G \int \frac{dm}{OP}
$$

$$
= -G \int \frac{dm}{\sqrt{R^2 + r^2 - 2Rr \cos \vartheta}}
$$

$$
= -\frac{G}{R} \int \left[1 - \frac{2r}{R} \cos \vartheta + \frac{r^2}{R^2} \right]^{-1/2} dm
$$

$$
= -\frac{G}{R} \int \left[1 + \frac{r}{R} \cos \vartheta \right.
$$

$$
\left. + (\tfrac{3}{2} \cos^2 \vartheta - \tfrac{1}{2}) \frac{r^2}{R^2} + O\left(\frac{r^3}{R^3} \right) \right] dm
$$

$$
= -\frac{G}{R} \left\{ \int dm + \frac{1}{R} \int r \cos \vartheta \, dm \right.
$$

$$
\left. + \frac{1}{R^2} \int (\tfrac{3}{2} \cos^2 \vartheta - \tfrac{1}{2}) r^2 \, dm + O\left(\frac{r^3}{R^3} \right) \right\}.
$$

The full expansion is effected through the introduction of the Legendre polynomials $P_n(\cos \vartheta)$. (See (562).)

The first integral gives the total mass M. Because G is at the center of mass, the second integral is zero. Let I be the moment of inertia of the body around the line GO. Then $I = \int r^2 \sin^2 \vartheta \, dm$. Moreover, the principal moments of inertia A, B, C satisfy

$$
(471)\qquad\qquad A + B + C = 2 \int r^2 \, dm.
$$

Therefore,

(472) $$\int(\tfrac{3}{2}\cos^2\vartheta - \tfrac{1}{2})r^2\,dm \;=\; \int(r^2 - \tfrac{3}{2}r^2\sin^2\vartheta)\,dm$$
$$=\; \tfrac{1}{2}(A + B + C) - \tfrac{3}{2}I.$$

Hence, up to terms of order r^3/R^3,

(473) $$V = -\frac{GM}{R} - \frac{G}{2R^3}(A + B + C - 3I).$$

This is *MacCullagh's formula*.

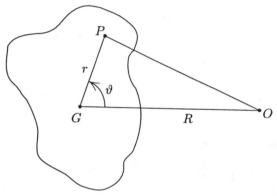

Figure VII.3. Point outside a body

Many celestial bodies are nearly spherical, both in geometry and in (apparent) distribution of mass. For them, A, B, C, and I are very nearly all equal, so the term in $1/R^3$ will be very small compared to the first term.

3. Precession of the equinoxes

Pre-Copernican, Graeco-Roman models of the cosmos put the Earth at the center of the universe, whose other objects were mounted on a system of concentric spherical, crystalline shells that rotated, one within the next, upon interpolated axes that necessarily pointed in many directions. As new phenomena of celestial motion were discovered, they were accommodated in the models by addition of the requisite number of new shells inserted into the nest. Naturally, each alteration at one layer called for recalculations of all shells outward—by Copernicus's time, nearly two hundred shells were called for.[2] Copernicus's technical innovation was to move the center of the nest of spheres from the center of the Earth to a point outside the Earth, cutting the number of shells needed drastically. The technical advantages of the move were swept away by Kepler's recognition of ellipses as the true planetary paths. The profound philosophical and theological consequences of course endure today.

Also enduring are commonplace phrases that 'should have' disappeared with the general spread of knowledge in the four centuries since Copernicus. For example, we still speak of the Sun 'rising' in the east and 'setting' in the west.[3] Phrases such 'celestial harmonies' testify to the extent that music theory and cosmology once were entwined.

One aspect of the Earth-centered cosmology remains useful today: The terminology is convenient for describing what can be seen from the Earth. We speak of the celestial sphere, which we imagine to be a sphere of very large radius, centered at the Earth's center and rotating on an axis that contains the Earth's rotation axis. Projection of the Earth's equator from the center onto the celestial sphere results in a great circle, the celestial equator, which divides the celestial sphere into northern and southern hemispheres. The apparent path of the Sun projects to a great circle called the ecliptic. The celestial equator and the ecliptic intersect at two points called the nodes; the point where the Sun moves from southern to northern is called the ascending node, the other the descending node. The times at which the Sun passes through the nodes are respectively the Spring and Fall equinoxes.

The ecliptic runs through a set of twelve star 'aggregations' or constellations which is called the zodiac. For historical reasons, the ascending node is sometimes called the First Point of Aries, even though the point now lies in the next constellation Pisces. In fact, the ascending node travels at a more-or-less steady rate around the ecliptic, completing a circuit in

---

[2]For a comprehensive look at the ultimate pre-Newtonian, philosophical developments of the universe of nested spheres, with interconnections to music—and to everything else!—see S.K. Heninger, Jr., *The Cosmographical Glass: Renaissance Diagrams of the Universe*, The Huntington Library, 1977, which is amply adorned with amazing illustrations.

[3]The names Levant (referring loosely to the countries east of the Mediterranean) and Europe seem etymologically to denote the rising and the setting, respectively, of the Sun.

about 25,800 years. This motion, called the precession of the equinoxes, was discovered by Hipparchus in about 128 B.C. At that time, the ascending node lay in Aries; some 2000 years earlier, in Taurus. In a few hundred years from now, the ascending node will pass into Aquarius.[4] According to a theory recently advanced by D. Ulansey,[5] the passing of the Spring equinox from Taurus into Aries may have furnished the basic symbolism for the Mithraic religion, which arose at the same time and in the same region as Christianity. Mithraism was carried throughout the Roman Empire by its armies and the religion became extinct in the fifth century A.D. at about the same time that the Roman Empire collapsed.

Now for the calculations. Recall MacCullagh's formula (473) for the potential of a rigid body of unit mass at a distant point: neglecting terms of higher order,

$$(474) \qquad V = -\frac{Gm}{R} - \frac{G}{2R^3}(A + B + C - 3I),$$

where R is the distance of the point from the center of mass of the body; A, B, C are the principal moments of inertia of the body; I is the moment of inertia around the line from the distant point through the center of mass; and m is the mass of the rigid body. The force per unit mass exerted by the body on the distant mass is $-\nabla V$. By Newton's Third Law, the distant mass exerts an equal and opposite force ∇V on the body, resulting in a torque per unit mass $r \times \nabla V$ around the origin at the center of mass, where r is the position of the distant point.

For a symmetric, homogeneous body, the quantity $A + B + C - 3I = 0$, and the vector product $r \times \nabla V = 0$. The 'real' Earth is not spherically homogeneous, and for it the product $r \times \nabla V \neq 0$. However, ∇V can be written as the sum of two terms, one of which is spherically symmetric and makes no contribution to the torque. Let the distant point be $r = (x, y, z)$,

---

[4] You may recall the 1960s musical *Hair*, which celebrated—albeit prematurely—'the dawning of the Age of Aquarius.'

[5] D. Ulansey, *The Origins of the Mithraic Mysteries: Cosmology and Salvation in the Ancient World*, Oxford University Press, 1989. D. Ulansey, 'Solving the Mithraic Mysteries,' *Biblical Archaeology Review*, **20**(September/October, 1994), p. 41.

a vector with direction cosines $(x/R, y/R, z/R)$. By (437),

(475) $$I = \frac{Ax^2 + By^2 + Cz^2}{R^3},$$

and we can write

(476) $$V = \text{spherically symmetric terms} + V_Q,$$

where

(477) $$V_Q = \frac{3GMI}{2R^3} = \frac{3GM}{2R^5}(Ax^2 + By^2 + Cz^2),$$

a potential energy and no longer per unit mass, contributes to the torque. Now it is easy to evaluate the torque $\mathbf{\Gamma} = \mathbf{r} \times \nabla V = \mathbf{r} \times \nabla V_Q$, with the result

(478) $$\mathbf{\Gamma} = \frac{3GM}{R^5}((C - B)yz, (A - C)zx, (B - A)xy).$$

EXERCISE VII.6. Verify (478) by carrying out the calculations.

We will describe the precession of the equinoxes by means of the Euler angles pictured in Figure VI.2. The two significant distant points exerting torques on the Earth are, of course, the Moon and the Sun. We are after a general prediction of the precession and we ignore many features of the real Earth-Moon-Sun system. We will also linearize the differential equations to be analyzed. Nevertheless, our conclusions agree with the 'average' behavior predicted by more detailed analyses.

The Earth, taken to be an oblate spheroid, possesses an axis of rotation around which it is symmetric and spins with angular velocity s. This angular velocity is also that of the moving frame along the principal axes of the Earth which rotates around the k-axis.[6] However, we may also introduce a rotating frame i, j, k whose axes do not rotate with the equatorial principal axes of the Earth but instead have angular velocity ω. Then the total angular velocity of the Earth will be $\omega + sk$, and the

---

[6]In fact, the principal axes in the equatorial plane are not uniquely determined: The moment of inertia of the Earth is essentially the same around every axis in the equatorial plane.

Euler equations of motion (445)—in vector form $\Gamma = (d/dt)(I\omega)$—when referred to the i, j, k frame becomes

$$
\begin{aligned}
(479) \qquad \Gamma_1 &= A\dot{\omega}_1 + (C - A)\omega_2\omega_3 + Cs\omega_2, \\
\Gamma_2 &= A\dot{\omega}_2 + (A - C)\omega_3\omega_1 - Cs\omega_1, \\
\Gamma_3 &= C(\dot{\omega}_3 + \dot{s}).
\end{aligned}
$$

Now suppose that the plane of i_0, j_0 in Figure VI.2 is the plane of the ecliptic, while the plane of i, j is the Earth's equatorial plane. The i_0, j_0, k_0 frame is inertial and the i, j, k frame is related to the Earth as already described. Pin down i now by requiring it to point along the intersection of the equatorial and ecliptic planes and toward the ascending node (the so-called *first point of Aries* and the position of the Spring equinox). Then the Euler angle ϑ, the angle between the vectors k_0 and k, is also the angle between the two planes. The Euler angle ψ is what we want to examine, because $d\psi/dt$ is the rate of precession. From Figure VI.2, we read off the components of ω:

$$
(480) \qquad \omega_1 = \frac{d\vartheta}{dt}, \qquad \omega_2 = \frac{d\psi}{dt}\sin\vartheta, \qquad \omega_3 = \frac{d\psi}{dt}\cos\vartheta.
$$

We note that our mild assumption that the Earth is an oblate spheroid, so that $A = B$, lets us simplify the expression (478) for Γ to

$$
(481) \qquad \Gamma = \frac{3GM}{R^5}((C - A)yz, -(C - A)zx, 0).
$$

EXERCISE VII.7. Show that (481) can also be written as

$$
(482) \qquad \Gamma = \frac{3GM}{R^5}(C - A)r{\cdot}k\, r\times k.
$$

Aiming to further simplify the torque (481), which must be transformed from rectangular coordinates (x, y, z) in the i, j, k frame into expressions in terms of the Euler angles ϑ, φ, ψ, we bring in the relative motions of the Earth and the distant point.

Suppose that the Earth revolves around the Sun (or that the Moon revolves around the Earth) in a circular path with constant angular velocity p. Because of our choice of the $i_0 j_0$-plane as the plane of the ecliptic, we can write the vector from the center of mass of the Earth to the distant point as

$$\text{(483)} \qquad \qquad \boldsymbol{r} = R(\cos pt\, \boldsymbol{i_0} + \sin pt\, \boldsymbol{j_0}),$$

where R is a constant and $\boldsymbol{i_0}, \boldsymbol{j_0}$ are fixed basis vectors of the inertial plane. To use the Euler equations, we must express (483) in terms of the frame $\boldsymbol{i}, \boldsymbol{j}, \boldsymbol{k}$. This is the same as working out the 'spherical' coordinates of the vectors $\boldsymbol{i_0}, \boldsymbol{j_0}$ on the unit sphere that is carried along with the frame $\boldsymbol{i}, \boldsymbol{j}, \boldsymbol{k}$. Thus,

$$\text{(484)} \qquad \begin{aligned} \boldsymbol{i_0} &= \cos\psi\, \boldsymbol{i} - \cos\vartheta \sin\psi\, \boldsymbol{j} + \sin\vartheta \sin\psi\, \boldsymbol{k}, \\ \boldsymbol{j_0} &= \sin\psi\, \boldsymbol{i} + \cos\vartheta \cos\psi\, \boldsymbol{j} - \sin\vartheta \cos\psi\, \boldsymbol{k}. \end{aligned}$$

From these and (483), we compute that

$$\text{(485)} \qquad \begin{aligned} \boldsymbol{r}\cdot\boldsymbol{k} &= -R\sin\vartheta \sin(pt - \psi), \\ \boldsymbol{r}\times\boldsymbol{k} &= R[\cos\vartheta \sin(pt - \psi)\, \boldsymbol{i} - \cos(pt - \psi)\, \boldsymbol{j}], \end{aligned}$$

and, using the result of Exercise VII.7, we find that the torque is

$$\text{(486)} \qquad \begin{aligned} \boldsymbol{\Gamma} = \ &\frac{3GM}{R^5}(C - A) \times \\ &[- \sin\vartheta \cos\vartheta \sin^2(pt - \psi)\, \boldsymbol{i} \\ &\quad + \sin\vartheta \sin(pt - \psi) \cos(pt - \psi)\, \boldsymbol{j}]. \end{aligned}$$

Now we can complete our analysis of the Euler equations (479). Because $\Gamma_3 = 0$, the third equation shows that $\omega_3 + s = \Omega$, a constant. Eliminating s from the first two Euler equations leads to the system

$$\text{(487)} \qquad \begin{aligned} A\dot{\omega}_1 - A\omega_2\omega_3 + C\Omega\omega_2 &= \Gamma_1, \\ A\dot{\omega}_2 - A\omega_3\omega_1 - C\Omega\omega_1 &= \Gamma_2. \end{aligned}$$

The angular velocities $\omega_1, \omega_2, \omega_3$ for the Earth are small compared to the angular velocity s, so also small compared to Ω. We therefore neglect the products $\omega_2\omega_3, \omega_3\omega_1$ compared to $\Omega\omega_2, \Omega\omega_1$. Using simple trigonometric identities, we then rewrite (487) as

$$(488) \quad A\dot{\omega}_1 + C\Omega\omega_2 = \frac{3GM}{2R^3}(C - A)\sin\vartheta\cos\vartheta[1 - \cos 2(pt - \psi)],$$

$$A\dot{\omega}_2 - C\Omega\omega_1 = \frac{3GM}{2R^3}(C - A)\sin\vartheta\sin 2(pt - \psi).$$

Because the precession is the motion of the vector i in the $i_0 j_0$-plane, ϑ is constant. The equations (488) show that ω_1, ω_2 are driven with angular frequency $2p$. The physical geometry of the Earth-Moon-Sun system indicates that ϑ is an acute angle and is roughly the same for both the Moon and the Sun because the Moon's path never strays far from the plane of the ecliptic. (Of course, the radii R are enormously different.) Furthermore, the angular accelerations $\dot{\omega}_1, \dot{\omega}_2$ are very small, so that we neglect $\dot{\omega}_1, \dot{\omega}_2$ with respect to $\Omega\omega_2, \Omega\omega_1$. Of course, $C \approx A$. Hence we obtain the determinations

$$(489) \quad \frac{d\psi}{dt} = \frac{\omega_2}{\sin\vartheta} = -\frac{3GM}{R^3\Omega}\frac{C - A}{C}\cos\vartheta[1 - \cos 2(pt - \psi)]$$

and

$$(490) \quad \frac{d\vartheta}{dt} = \omega_1 = -\frac{3GM}{R^3\Omega}\frac{C - A}{C}\sin\vartheta\sin 2(pt - \psi).$$

The full precession of the equinoxes, as observed from the time of Hipparchus (ca. 128 B.C.) until the present, takes about 25,800 years, while the periods of the Sun and the Moon in their apparent motions around the Earth are 1 year and about $1/12$ year, respectively. We may suppose therefore that ψ remains constant during a year or portion thereof. The average of $1 - \cos 2(pt - \psi)$ over the yearly period $T = \pi/p$ is then

$$(491) \quad \frac{1}{T}\int_0^T [1 - \cos 2(pt - \psi)]\, dt = 1.$$

We can say then that the precession proceeds with average yearly rate

$$(492) \quad \overline{d\psi/dt} = -\frac{3GM}{2R^3\Omega}\frac{C - A}{C}\cos\vartheta,$$

and, because $C > A$ and there is an additional multiplier -1, in the sense opposite to the sense of the Earth's rotation; that is, the precession moves from east to west. This is why the Spring equinox proceeds 'backwards' through the zodiac: from Taurus into Aries into Pisces into Aquarius into ... .

The joint precessional effect of the Moon and the Sun is very nearly the sum of the separate effects because the orbit of the Moon lies very close to the plane of the ecliptic. The observed precession of the equinoxes leads to a calculated value for the constant $(C - A)/C$. In recent years, this constant has been calculated to a high precision because of the opportunity to make long-term observations of orbital precessions of hundreds of artificial Earth satellites.

Before leaving this topic, we must make right a little fib. It is clear from (490) that ϑ is not constant but has a slight drift of its own. Because the average value of $\sin 2(pt - \pi)$ will be essentially 0, ϑ will 'wobble' back and forth across a 'normal' value, which for the Earth is about 23.5°. This motion is called *nutation,* Latin for 'nodding,' and the nutation of the Earth is caused principally by the Moon with a period of 18.6 years. Nutation is most easily seen in the motion of a 'sleeping' top which has settled into steady precession. Gravity causes the sleeping top to nod.

EXERCISE VII.8. Obtain a top and set it spinning. Observe the precession and look for nutation.

4. Internal potential of a homogeneous ellipsoid

There is no simple way known to calculate the potential of a homogeneous ellipsoid. All ways require some nasty integrations. We are interested in both the internal and the external fields. The calculation is done in several stages.

First, we will find the internal potential of a homoid. We calculated on page 124 that the internal gravitational force is zero; therefore, the potential is constant and everywhere equal to its value at the center O. Let the outer ellipsoid have axes a, b, c, and the inner ellipsoid, axes pa, pb, pc, where $p < 1$. Take a cone of solid angle $d\omega$ and vertex O, intersecting the boundaries at distances pr and r from O. The potential at O due to the frustum intercepted by this cone is

(493) $$-G \int_{pr}^{r} \rho r^2 \, d\omega \, \frac{dr}{r} = -\tfrac{1}{2} G \rho (1 - p^2) r^2 \, d\omega.$$

Therefore, the total potential at O is

(494) $$V_O = -\tfrac{1}{2} G \rho (1 - p^2) \int r^2 \, d\omega,$$

the integral being extended over the outer boundary.

The equation of the outer boundary is

(495) $$\frac{x^2}{a^2} + \frac{y^2}{b^2} + \frac{z^2}{c^2} = 1.$$

Let a point on the outer boundary have coordinates (rl, rm, rn), where l, m, n are direction cosines ($l^2 + m^2 + n^2 = 1$). Then

(496) $$r^2 \left(\frac{l^2}{a^2} + \frac{m^2}{b^2} + \frac{n^2}{c^2} \right) = 1.$$

Take spherical coordinates r, ϑ, φ such that

(497) $$l = \sin \vartheta \cos \varphi, \quad m = \sin \vartheta \sin \varphi, \quad n = \cos \vartheta.$$

Then the element of solid angle is $d\omega = \sin \vartheta \, d\vartheta \, d\varphi$.

The integrand in the expression (494) for V_O is symmetric in each axis. Therefore, its integrated value is eight times the value obtained by integrating over the first octant. Hence

(498)
$$\begin{aligned}
V_O &= -\tfrac{1}{2} G \rho (1 - p^2) \times \\
&\quad \int \frac{d\omega}{l^2/a^2 + m^2/b^2 + n^2/c^2} \\
&= -4 G \rho (1 - p^2) \times \\
&\quad \int_0^{\pi/2} \int_0^{\pi/2} \frac{\sin \vartheta \, d\vartheta \, d\varphi}{\sin^2 \vartheta (\cos^2 \varphi/a^2 + \sin^2 \varphi/b^2) + \cos^2 \vartheta/c^2}.
\end{aligned}$$

We know that

(499) $$\int_0^{\infty} \frac{dt}{A^2 + B^2 t^2} = \left[\frac{1}{AB} \tan^{-1} \left(\frac{Bt}{A} \right) \right]_0^{\infty} = \frac{\pi}{2AB}.$$

Let $\tan \varphi = t$. Then

$$
\begin{aligned}
V_O \; &= \; -4G\rho(1 - p^2) \times \\
(500) \quad & \int_0^\infty \int_0^{\pi/2} \frac{\sin \vartheta \, d\vartheta \, dt}{\sin^2 \vartheta/a^2 + \cos^2 \vartheta/b^2 + (\sin^2 \vartheta/b^2 + \cos^2 \vartheta/c^2)t^2} \\
&= \; -2\pi G\rho(1 - p^2) \times \\
& \int_0^{\pi/2} \frac{\sin \vartheta \, d\vartheta}{[(\sin^2 \vartheta/a^2 + \cos^2 \vartheta/c^2)(\sin^2 \vartheta/b^2 + \cos^2 \vartheta/c^2)]^{1/2}} \, .
\end{aligned}
$$

Putting $u = c^2 \tan^2 \vartheta$, substituting, and rearranging, we obtain

$$(501) \qquad\qquad V_O = -\pi G\rho(1 - p^2)abcI,$$

where

$$(502) \qquad\qquad I = \int_0^\infty \frac{du}{\Delta}$$

and

$$(503) \qquad\qquad \Delta = \sqrt{(a^2 + u)(b^2 + u)(c^2 + u)}.$$

Note that the expression (501) for V_O is, as it should be, symmetric in a, b, c. If a, b, c are pairwise distinct, then the integral (502) for I is an elliptic integral. If two of a, b, c are equal, then the integral can be evaluated in terms of elementary functions, as we shall do later. If $p = 0$, we obtain the potential at the center of a homogeneous ellipsoid.

Next, we find the components of the attractive force at a point $P(x, y, z)$ inside a solid and homogeneous ellipsoid. Because the homoid outside the similar ellipsoid passing through P exerts no force at P (page 124), it is sufficient to find the force exerted by the ellipsoid at a point on its surface.

Let a line from P in the direction (l, m, n) meet the surface at Q, where $PQ = r$. The coordinates of Q are then $(x + lr, y + mr, z + nr)$. The conditions for P and Q to lie on the ellipsoid with axes a, b, c are given by equation (495) and

$$(504) \qquad\qquad \frac{(x + lr)^2}{a^2} + \frac{(y + mr)^2}{b^2} + \frac{(z + nr)^2}{c^2} = 1.$$

By subtraction, we find the condition

$$(505) \qquad r^2 \left(\frac{l^2}{a^2} + \frac{m^2}{b^2} + \frac{n^2}{c^2} \right) + 2r \left(\frac{lx}{a^2} + \frac{my}{b^2} + \frac{nz}{c^2} \right) = 0.$$

The root $r = 0$ gives P; the other root gives Q.

Consider the cone with vertex P, solid angle $d\omega$, and axis along PQ. The force of attraction at P due to the matter in this cone is

$$(506) \qquad -G\rho \, d\omega \int_0^r \frac{r^2 \, dr}{r^2} = -G\rho r \, d\omega.$$

Therefore, the components of this force at P are

$$(507) \qquad (-G\rho lr \, d\omega, -G\rho mr \, d\omega, -G\rho nr \, d\omega).$$

To get the total components of force, we integrate over all ω. But (l, m, n) determines a *ray* from P, while ω determines a *line*. Therefore, the integral gives double the value sought. The x-component of the total force is

$$
\begin{aligned}
(508) \qquad X &= -\tfrac{1}{2} G\rho \int lr \, d\omega \\
&= -\tfrac{1}{2} G\rho \int \frac{2l(lx/a^2 + my/b^2 + nz/c^2)}{l^2/a^2 + m^2/b^2 + n^2/c^2} \, d\omega.
\end{aligned}
$$

From the symmetry across coordinate planes, integrals of products lm, mn, nl over the sphere of directions are zero; the same is true if such products are multiplied by even functions of l, m, n. Hence

$$
\begin{aligned}
(509) \qquad X &= -G\rho x \int \frac{l^2/a^2}{l^2/a^2 + m^2/b^2 + n^2/c^2} \, d\omega \\
&= -G\rho x A,
\end{aligned}
$$

where A denotes the integral.[7]

---

[7] There is some danger of confusing A (as used here for the integral) with A (as used for a principal moment of inertia). When we consider figures of equilibrium, we will take care. The notation is standard.

The integral for A is reduced by the same methods used to arrive at (501). Thus, using spherical polar coordinates and setting $t = \tan\varphi$,

$$A = 8 \int_0^{\pi/2} \int_0^{\pi/2} \frac{(\cos^2\vartheta/a^2)\sin\vartheta \, d\vartheta \, d\varphi}{\cos^2\vartheta/a^2 + \sin^2\vartheta(\cos^2\varphi/b^2 + \sin^2\varphi/c^2)}$$

$$(510) = 8 \int_0^\infty \int_0^{\pi/2} \frac{(\cos^2\vartheta/a^2)\sin\vartheta \, d\vartheta \, dt}{\cos^2\vartheta/a^2 + \sin^2\vartheta/b^2 + (\cos^2\varphi/b^2 + \sin^2\varphi/c^2)t^2}$$

$$= 4\pi \int_0^{\pi/2} \frac{(\cos^2\vartheta/a^2)\sin\vartheta \, d\vartheta}{\sqrt{(\cos^2\vartheta/a^2 + \sin^2\vartheta/b^2)(\cos^2\varphi/a^2 + \sin^2\varphi/c^2)}}.$$

Put $u = a^2 \tan^2\vartheta$; after some reduction we obtain

$$(511) \qquad\qquad A = 2\pi abc \int_0^\infty \frac{du}{(a^2 + u)\Delta}.$$

This is again an elliptic integral if a, b, c are all distinct. There are symmetrical expressions for Y and Z in terms of integrals B and C.

Now we can find the potential at any point inside a homogeneous ellipsoid. The gravitational force

$$(512) \qquad\qquad (X, Y, Z) = (-G\rho Ax, -G\rho By, -G\rho Cz).$$

Therefore, we can find a constant D such that

$$(513) \qquad\qquad V = -\tfrac{1}{2}G\rho(D - Ax^2 - By^2 - Cz^2).$$

To evaluate D, set $x = y = z = 0$. From (501), the potential at the center of a homogeneous ellipsoid is $V_O = -\pi G\rho abcI$, whereas we have identified $V_O = -\tfrac{1}{2}G\rho D$. Thus, $D = 2\pi abcI$. Combining all of the expressions for D, then the integral representations of A, B, C, we find that

$$(514)\,V = -G\pi\rho abc \int_0^\infty \left\{ 1 - \frac{x^2}{a^2 + u} - \frac{y^2}{b^2 + u} - \frac{z^2}{c^2 + u} \right\} \frac{du}{\Delta}$$

$$= -G\pi\rho abc \left\{ 1 + \frac{x^2}{a} \frac{\partial I}{\partial a} + \frac{y^2}{b} \frac{\partial I}{\partial b} + \frac{z^2}{c} \frac{\partial I}{\partial c} \right\}.$$

It will prove useful later to note now that

$$(515) \qquad\qquad A + B + C = \int d\omega = 4\pi.$$

Also,

$$(516) \qquad Aa^2 + Bb^2 + Cc^2 = \int \frac{d\omega}{l^2/a^2 + m^2/b^2 + n^2/c^2}.$$

Note finally that A, B, C depend only upon the *ratios* of the axes.

As we have mentioned above, the integrals for A, B, C, and I can in general be evaluated in terms of elliptic integrals and elliptic functions. For *spheroids,* two of the axes are equal, and the integrals can be evaluated in terms of elementary functions.

Let $a = b$, so that $A = B$. From (516), setting $u = \cos \vartheta$,

$$
\begin{aligned}
(517) \qquad 2Aa^2 + Cc^2 &= \int \frac{d\omega}{(l^2 + m^2)/a^2 + n^2/c^2} \\
&= \int_0^{2\pi} \int_0^{\pi} \frac{\sin \vartheta \, d\vartheta \, d\varphi}{\sin^2 \vartheta/a^2 + \cos^2 \vartheta/c^2} \\
&= 2\pi \int_{-1}^{1} \frac{du}{1/a^2 + (1/c^2 - 1/a^2)u^2}.
\end{aligned}
$$

There are two cases.

CASE 1. Oblate spheroid, $a > c$. Then

$$
\begin{aligned}
(518) \qquad 2Aa^2 + Cc^2 &= \frac{4\pi a^2 c}{\sqrt{a^2 - c^2}} \tan^{-1} \sqrt{\frac{a^2}{c^2} - 1} \\
&= 4\pi a^2 \frac{\sqrt{1 - e^2}}{e} \sin^{-1} e,
\end{aligned}
$$

where $e = \sqrt{a^2 - c^2}/a$ is the eccentricity of the meridian ellipse.

CASE 2. Prolate spheroid, $a < c$. Then

$$
\begin{aligned}
(519) \qquad 2Aa^2 + Cc^2 &= \frac{2\pi a^2 c}{\sqrt{c^2 - a^2}} \log \left\{ \frac{c + \sqrt{c^2 - a^2}}{c - \sqrt{c^2 - a^2}} \right\} \\
&= 4\pi c^2 \frac{\sqrt{1 - e^2}}{e} \tanh^{-1} e,
\end{aligned}
$$

where $e = \sqrt{c^2 - a^2}/c$ is the eccentricity of the meridian ellipse.

The separate values for A and C can be found by noting that $2A + C = 4\pi$ and solving a pair of linear equations for A and C.

We will apply these formulas to the case when the spheroids or ellipsoids differ only slightly from spheres. The *flattening* of an ellipse with axes

$a > c$ is

(520)
$$\epsilon = \frac{a - c}{a}.$$

The flattening and the eccentricity are related by the equation

(521)
$$e^2 = 2\epsilon + \epsilon^2.$$

If e and ϵ are both small, then $e^2 \cong 2\epsilon$. For the rest of this section, we will assume that e and ϵ are small.

In Case 1, the oblate spheroid, note that

(522)
$$
\begin{aligned}
\frac{\sqrt{1 - e^2}}{e} \sin^{-1} e \; &= \; \sqrt{1 - e^2} \cdot \frac{\sin^{-1} e}{e} \\
&= \; \left(1 - \tfrac{1}{2}e^2 - \tfrac{1}{8}e^4 + \cdots\right)\left(1 + \tfrac{1}{6}e^2 + \tfrac{3}{40}e^4 + \cdots\right) \\
&= \; 1 - \tfrac{1}{3}e^2 - \tfrac{2}{15}e^4 + \cdots .
\end{aligned}
$$

Moreover, $c^2 = a^2(1 - e^2)$, so that

(523)
$$2A + C(1 - e^2) = 4\pi[1 - \tfrac{1}{3}e^2 - \tfrac{2}{15}e^4 + O(e^6)].$$

Also,

(524)
$$2A + C = 4\pi.$$

Because $e^2 \cong 2\epsilon$, we calculate that

(525)
$$
\begin{aligned}
A &= \tfrac{4}{3}\pi(1 - \tfrac{2}{5}\epsilon + \cdots), \\
C &= \tfrac{4}{3}\pi(1 + \tfrac{4}{5}\epsilon + \cdots).
\end{aligned}
$$

In Case 2, the prolate spheroid, we have $c > a$. Then

(526)
$$
\begin{aligned}
A &= \tfrac{4}{3}\pi(1 + \tfrac{2}{5}\epsilon + \cdots), \\
C &= \tfrac{4}{3}\pi(1 - \tfrac{4}{5}\epsilon + \cdots),
\end{aligned}
$$

where $a = c(1 - \epsilon)$.

Now consider the case of an ellipsoid with axes $a, a(1 - \epsilon), a(1 - \eta)$, there being two flattenings. Because higher powers of ϵ and η are being neglected, A, B, C will be approximated by functions linear in ϵ and η. When $\eta = 0$, we have found above that—with a little renaming—$A =$

$B = \frac{4}{3}\pi(1-\frac{2}{5}\epsilon)$ and $C = \frac{4}{3}\pi(1+\frac{4}{5}\epsilon)$. When $\epsilon = 0$, $A = C = \frac{4}{3}\pi(1-\frac{2}{5}\eta)$ and $B = \frac{4}{3}\pi(1+\frac{4}{5}\eta)$. Therefore, *in general,*

$$
(527) \qquad
\begin{aligned}
A &= \tfrac{4}{3}\pi(1 - \tfrac{2}{5}\epsilon - \tfrac{2}{5}\eta), \\
B &= \tfrac{4}{3}\pi(1 - \tfrac{2}{5}\epsilon + \tfrac{4}{5}\eta), \\
C &= \tfrac{4}{3}\pi(1 + \tfrac{4}{5}\epsilon - \tfrac{2}{5}\eta).
\end{aligned}
$$

These formulas can be rendered symmetrical by introducing the *mean radius* $k = \frac{1}{3}(a + b + c)$. Then

$$(528) \qquad A = \tfrac{4}{3}\pi\left(1 - \tfrac{6}{5}\tfrac{a-k}{k}\right),$$

$$(529) \qquad B = \tfrac{4}{3}\pi\left(1 - \tfrac{6}{5}\tfrac{b-k}{k}\right),$$

$$(530) \qquad C = \tfrac{4}{3}\pi\left(1 - \tfrac{6}{5}\tfrac{c-k}{k}\right).$$

EXERCISE VII.9. Show that, if $a > b > c$, then $Aa^2 > Bb^2 > Cc^2$; and that at points on the surface of the ellipsoid the potential is greatest at the end of the axis c.

EXERCISE VII.10. Prove that a spheroid of uniform density can not have its boundary surface as an equipotential surface.

EXERCISE VII.11. Prove that the equipotentials inside a solid, homogeneous ellipsoid are similar and similarly situated ellipsoids.

EXERCISE VII.12. Find the amount by which the gravitational potential of a uniform, solid ellipsoid exceeds that of a uniform sphere of equal volume and mass.

EXERCISE VII.13. Show that for a nearly spherical ellipsoid for which

$$(531) \qquad a = k(1 + \lambda), \quad b = k(1 + \mu), \quad c = k(1 + \nu),$$

where λ, μ, ν are small, $k^3 = abc$, and $\lambda + \mu + \nu = 0$ approximately, the components of gravitational attraction at an internal point (x, y, z) are

$$(532) \qquad \left(-\tfrac{4}{3}\pi G\rho(1 - \tfrac{6}{5}\lambda)x, \ -\tfrac{4}{3}\pi G\rho(1 - \tfrac{6}{5}\mu)y, \ -\tfrac{4}{3}\pi G\rho(1 - \tfrac{6}{5}\nu)z\right)$$

EXERCISE VII.14. For a prolate spheroid with $a = b = c\sqrt{1 - e^2}$, prove that

$$(533) \qquad C = \frac{4\pi}{3}\left\{1 - 6\sum_{n=1}^{\infty}\frac{e^{2n}}{(2n+1)(2n+3)}\right\}.$$

5. External potential of a homogeneous ellipsoid

Now we turn to the external potential of a homogeneous ellipsoid. No way has been discovered to calculate this function directly. It will be convenient to proceed by way of a theorem due to Ivory.

Let the boundary S of the ellipsoid be given by the equation

(534)
$$\frac{x^2}{a^2} + \frac{y^2}{b^2} + \frac{z^2}{c^2} = 1.$$

An ellipsoid $S(\lambda)$ with the equation

(535)
$$\frac{x^2}{a^2 + \lambda} + \frac{y^2}{b^2 + \lambda} + \frac{z^2}{c^2 + \lambda} = 1$$

is said to be *confocal* with $S = S(0)$. If (x, y, z) is considered to be a fixed point on the surface, then the equation (535) is a cubic equation to be satisfied by λ. Suppose that $a > b > c$. Then this cubic equation has exactly one positive root. To see this, begin by setting

(536)
$$\varphi(\lambda) = \frac{x^2}{a^2 + \lambda} + \frac{y^2}{b^2 + \lambda} + \frac{z^2}{c^2 + \lambda} - 1.$$

A little reflection will lead to the graph of $\varphi(\lambda)$ as depicted in Figure VII.4.

There is one root in each of the intervals $(-a^2, -b^2)$, $(-b^2, -c^2)$, and $(-c^2, \infty)$. The corresponding quadric surfaces will be respectively a hyperboloid of two sheets, a hyperboloid of one sheet, and an ellipsoid.[8]

Let $P'(x', y', z')$ be a point on $S(\lambda)$ and define $P(x, y, z)$ by

(537)
$$\frac{x}{a} = \frac{x'}{a'}, \quad \frac{y}{b} = \frac{y'}{b'}, \quad \frac{z}{c} = \frac{z'}{c'},$$

where $a'^2 = a^2 + \lambda$, etc. Then P must lie on S. The points P and P' are called *corresponding points,* and $P' \leftrightarrow P$ establishes a one-to-one correspondence between $S(\lambda)$ and S.

---

[8]These three surfaces intersect orthogonally at (x, y, z). The three roots (λ, μ, ν) are called the *confocal coordinates* of (x, y, z).

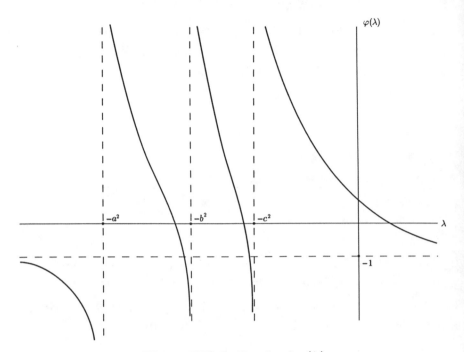

Figure VII.4. Graph of $\varphi(\lambda)$

Let QR be an elementary strip of S with cross-section $dy\,dz$, parallel to the x-axis, and let $Q'R'$ be the corresponding strip of $S(\lambda)$ with cross-section $dy'\,dz'$ (see Figure VII.5), so that

$$(538) \qquad \frac{dy\,dz}{dy'\,dz'} = \frac{bc}{b'c'}.$$

If $f'(r)$ denotes the magnitude of force at distance r (inverse square for Newtonian attraction) and if ρ is the density function, then the attraction at P' due to the strip QR has as its x-component the quantity

$$(539) \qquad \Delta X = -G\rho\,dy\,dz \int f'(r)\cos(\angle P'TR)\,dx,$$

where T is the position of the volume element $dx\,dy\,dz$. But $x^2+y^2+z^2 = r^2$, r being the distance from the origin temporarily at P'. If y and z are constant, as they are along each strip, then $x\,dx = r\,dr$, so that

$$(540) \qquad \cos(\angle P'TR) = \frac{x}{r} = \frac{dr}{dx}.$$

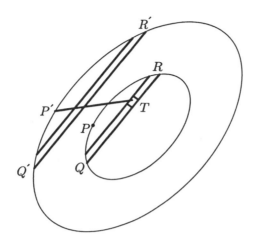

Figure VII.5. Elementary strips

Hence,

(541)
$$\Delta X = -G\rho \, dy \, dz \int \frac{df}{dr} \frac{dr}{dx} \, dx$$
$$= -G\rho \, dy \, dz \, [f(P'R) - f(P'Q)].$$

Similarly, the x-component of the gravitational attraction at P due to the strip $Q'R'$ is

(542)
$$\Delta X' = -G\rho \, dy' \, dz' \, [f(PR') - f(PQ')].$$

From the definition of corresponding points, we conclude—by the confocality of S and $S(\lambda)$—that $PQ' = P'Q$ and $PR' = P'R$ for any pair of corresponding points. Taking the ratio of forces, we get $\Delta X / \Delta X' = dy \, dz / dy' \, dz' = bc/b'c'$. Integrating over all strips QR of S and $Q'R'$ of $S(\lambda)$ that are in correspondence, we have finally $X/X' = bc/b'c'$. This is

> *Ivory's Theorem:* If S and $S(\lambda)$ are confocal ellipsoids with corresponding points P and P' on their surfaces, and if X is the x-component of attraction of the ellipsoid S

at P' and X' that of $S(\lambda)$ at P, then

(543)
$$\frac{X}{X'} = \frac{bc}{b'c'}.$$

Note that the conclusion of Ivory's Theorem holds true for any law of force that depends only upon distance.

Now, P lies inside $S(\lambda)$. Therefore, we can apply Ivory's Theorem with S the ellipsoid confocal to $S(\lambda)$ whose surface passes through P. By (509), $X' = -A'G\rho x$, where A' is the same function of a', b', c that A is of a, b, c. Therefore,

(544)
$$X = -\frac{bc}{b'c'}A'G\rho x = -\frac{abc}{a'b'c'}A'G\rho x',$$

with similar expressions for Y and Z.

To calculate the components of force at P', it is necessary to find from (x', y', z') the appropriate value of λ; then (a', b', c'); and then (A', B', C'). Then

(545)
$$X = -\frac{abc}{a'b'c'}G\rho x' \int_0^\infty \frac{2\pi a'b'c'\, du}{(a'^2 + u)\Delta'(u)},$$

where $\Delta'(u)$ is the same function of a', b', c', u that $\Delta(u)$ is of a, b, c, u. Put $u = v - \lambda$. Then

(546)
$$X = -2\pi G\rho abc x' \int_\lambda^\infty \frac{dv}{(a^2 + v)\Delta(v)},$$

with similar expressions for Y and Z.

These expressions for the forces X, Y, Z suggest that the exterior potential might be

(547)
$$V = -\pi G\rho abc \int_\lambda^\infty \left\{ 1 - \frac{x'^2}{a^2 + v} - \frac{y'^2}{b^2 + v} - \frac{z'^2}{c^2 + v} \right\} \frac{dv}{\Delta(v)}.$$

By symmetry, we need check only that $\partial V / \partial x = -X$. By direct calculation, remembering that λ is a function of (x', y', z'), we get

(548)
$$-\frac{\partial V}{\partial x} = -2\pi G\rho abc x' \int_\lambda^\infty \frac{dv}{(a^2 + \lambda)\Delta(v)}$$
$$- \pi G\rho abc \frac{\partial \lambda}{\partial x} \left\{ 1 - \frac{x'^2}{a^2 + v} - \frac{y'^2}{b^2 + v} - \frac{z'^2}{c^2 + v} \right\} \frac{1}{\Delta(\lambda)}.$$

Because (x', y', z') is on $S(\lambda)$, the curly-bracketed term $\{\;\} = 0$, and only the first term, which is X, remains. Moreover, the additive constant is appropriate: if (x', y', z') is on S, then $\lambda = 0$. Compare (547) with the expression (514) to verify that the potential V is continuous across the surface S.

We conclude this section by deducing the gravitational attraction toward an infinitely long cylinder having an elliptical cross-section. Let the ellipse have semiaxes $a > b$. Consider an ellipsoid with axes a, b, c. The components of attraction at the point $(x, y, 0)$ in the equatorial plane can be obtained from the formulas already found.

INTERNAL CASE. The attracting force is $(-G\rho Ax, -G\rho By, 0)$ by (512). Pass from the ellipsoid to the cylinder by letting $c \to \infty$. Then A turns into

$$(549) \qquad A = 2\pi ab \int_0^\infty \frac{du}{(a^2 + u)\sqrt{(a^2 + u)(b^2 + u)}},$$

an 'elementary' integral. Let $v^2 = (a^2 + u)^{-1}$; after some manipulation, the integral becomes

$$(550) \qquad A = 4\pi ab \int_0^{1/a} \frac{dv}{\sqrt{1 - v^2(a^2 - b^2)}} = \frac{4\pi b}{a + b}.$$

Similarly,

$$(551) \qquad B = \frac{4\pi a}{a + b}.$$

Therefore,

$$(552) \qquad X = -\frac{4\pi \rho ab}{a + b}\frac{x}{a} \quad \text{and} \quad Y = -\frac{4\pi \rho ab}{a + b}\frac{y}{b}.$$

EXTERNAL CASE. Using the expressions (546), we find in a similar way that, at an external point (x, y),

$$(553) \qquad X = -\frac{4\pi \rho ab}{a' + b'}\frac{x}{a'} \quad \text{and} \quad Y = -\frac{4\pi \rho ab}{a' + b'}\frac{y}{b'},$$

where a' and b' are the semiaxes of an ellipse confocal with the cross-section of the cylinder and passing through (x, y).

VIII

Shape of a Self-Gravitating Fluid

1. Hydrostatic equilibrium

Consider a mass of liquid, incompressible but not necessarily homogeneous in density, that rotates around a fixed axis through its center of mass without any external force. Take the z-axis to be the rotation axis, and assume that the angular velocity is the constant ω. The x- and y-axes will be fixed in the fluid mass with the origin at the center of gravity. Denote by p the pressure at a point (x, y, z) of the fluid; the pressure depends only upon the position. The force acting upon an element of volume $d\tau$ is

$$(554) \qquad (X \, d\tau, Y \, d\tau, Z \, d\tau) = \left(\frac{\partial p}{\partial x} \, d\tau, \frac{\partial p}{\partial y} \, d\tau, \frac{\partial p}{\partial z} \, d\tau \right),$$

where (X, Y, Z) is the force per unit volume at (x, y, z). The equations of motion can be written in the rotating xyz-coordinate system by including a 'fictitious' centripetal force. Written per unit volume, the equations of motion of the fluid are expressed in terms of the potential energy V by

$$(555) \qquad \rho \frac{\partial V}{\partial x} = X - \omega^2 \rho x,$$

$$(556) \qquad \rho \frac{\partial V}{\partial y} = Y - \omega^2 \rho y,$$

$$(557) \qquad \rho \frac{\partial V}{\partial x} = Z.$$

The potential modified for rotation (page 15) is

(558) $$U = V - \tfrac{1}{2}\omega^2(x^2 + y^2) \qquad \text{(per unit mass)}.$$

Thus, the condition of equilibrium is

(559) $$\left(\frac{\partial p}{\partial x}, \frac{\partial p}{\partial y}, \frac{\partial p}{\partial z}\right) = \left(\rho\frac{\partial U}{\partial x}, \rho\frac{\partial U}{\partial y}, \rho\frac{\partial U}{\partial z}\right).$$

The equilibrium equation can be written more concisely in either of the two forms

(560) $$\nabla p = \rho \nabla U \quad \text{or} \quad dp = \rho\, dU.$$

If $dU = 0$, then $dp = 0$; therefore, p is a function only of U. The surfaces $p = $ constant and $U = $ constant coincide. The surface $U = $ constant is a *level surface* or *equipotential*. Consequently, a level surface is a surface of equal pressure (an *isobaric surface*). It follows from $dp = \rho\, dU$ that ρ is then a function of U, so that a level surface is also a surface of constant density (an *isopycnic surface*). There can be no forces acting on the surface S of the fluid mass if the mass is in equilibrium. Therefore, $p = 0$ on S, whence U is constant on S. The free surface S is a level surface with an equation $U = $ constant. The force of gravity is then $-\nabla U$, normal to the surface.

2. Distortion of a liquid sphere by a distant mass

We apply the formulas obtained in the previous chapter for the internal potential of a homogeneous ellipsoid to study the tide-raising action of a distant body on a liquid mass. For the present, we consider a spherical mass M of a homogeneous fluid of density ρ, supposed not to be rotating. (It will become clear later how to include the effects of rotation.) This mass, the *primary*, will be subject to the gravitational influence of a second body situated at a great distance. If that body is nearly spherically symmetrical, MacCullagh's formula (473) implies that the distant body very nearly has the gravitational effect of a point mass M'.

Take the origin at the center of mass of the primary, and introduce spherical polar coordinates (r, ϑ, φ), where φ is longitude and ϑ is colatitude with respect to the axis OM. (See Figure VIII.1.) The potential at (r, ϑ, φ) due to M' at the point $(R, 0, \varphi)$, where $r \ll R$, is given by

(561)
$$V = -\frac{GM'}{\sqrt{R^2 + r^2 - 2Rr \cos \vartheta}} = -\frac{GM'}{R} \frac{1}{\sqrt{1 - 2r/R \cos \vartheta + r^2/R^2}}.$$

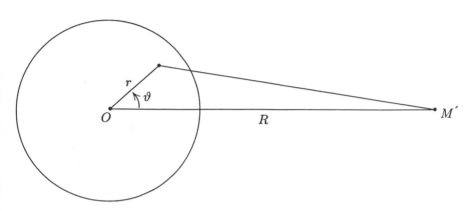

Figure VIII.1. Sphere and distant mass

By assumption, $r/R \ll 1$. Therefore, the factor with radical can be expanded into a series in powers of r/R:

(562)
$$\frac{1}{\sqrt{1 - 2r/R \cos \vartheta + r^2/R^2}} = \sum_{n=0}^{\infty} P_n(\cos \vartheta) \left(\frac{r}{R}\right)^n.$$

In this expansion, carried out by means of the binomial series, the coefficient $P_n(x)$ is easily seen to be a polynomial of degree n in x: it is the nth *Legendre polynomial*. The first few cases are $P_0(x) = 1$, $P_1(x) = x$, $P_2(x) = \frac{1}{2}(3x^2 - 1)$, $P_3(x) = \frac{1}{2}(5x^3 - 3x)$, $P_4(x) = \frac{1}{8}(35x^4 - 3x^2 + 3)$.[1] (See Exercise VIII.1.)

Therefore,

(564) $V = -\dfrac{GM'}{R} - \dfrac{GM'r \cos \vartheta}{R^2} - \dfrac{GM'r^2(3 \cos^2 \vartheta - 1)}{2R^3} + \cdots .$

---

[1] In general,

(563) $P_n(x) = \dfrac{(2n)!}{2^n (n!)^2} \left\{ x^n - \dfrac{n(n-1)}{2(2n-1)} x^{n-2} + \dfrac{n(n-1)(n-2)(n-3)}{2 \cdot 4(2n-1)(2n-3)} x^{n-4} + \cdots \right\},$

the series ending in a constant if n is even, in an x^1-term if n is odd.

The first term of this expansion is constant, so it engenders no tide-raising force. The second term is $-GM'x/R^2$, where x is a coordinate along the line of centers OM'. This gives a constant force GM'/R^2 toward M' from O. This force can be discarded if at the same time the system is rotating about an axis perpendicular to OM' with angular velocity ω satisfying $R\omega^2 = GM'/R^2$ while the center of mass of M remains at the origin. This condition is always fulfilled in the Earth-Moon system to high accuracy, for the mass of the Moon is about $1/81$ of the mass of the Earth, and the eccentricity of the orbit of the Moon is about $1/20$, so that the orbit may be approximated for our needs by a circle. Moreover, the condition on ω is just Kepler's Third Law, considering the Earth to be in a circular orbit around the Moon!

The remaining terms in the expansion (561) constitute the *tide-raising potential* V_T. Because $r/R \ll 1$, we consider only the first term of V_T. In Cartesian coordinates, where the x-axis points toward M' and the z-axis is the axis of rotation, we have[2]

$$(565) \qquad V_T = K(x^2 - \tfrac{1}{2}y^2 - \tfrac{1}{2}z^2),$$

where $K = -GM'/R^3$ is a constant. The goal is to find the effect of V_T on the shape of the primary. The pertinent condition is that the free surface of M must be an equipotential in the combined fields produced by M and M'. If this were not so, there would be places on the free surface where $dU \neq 0$. Because $\rho \neq 0$, there would be places where $dp \neq 0$, so that there would be pressure on the surface. Then the surface would move, whereas we are assuming it to be in equilibrium.

Let us assume that the equilibrium configuration of the surface of M is the ellipsoid with equation

$$(566) \qquad \frac{x^2}{a^2} + \frac{y^2}{b^2} + \frac{z^2}{c^2} = 1.$$

The potential at (x, y, z) within M or on the surface is compounded of two parts: that due to M' and just computed, and the internal potential (514).

---

[2]Remember that $r^2 \cos^2 \vartheta = x^2$ and $r^2 = x^2 + y^2 + z^2$.

Thus,

(567) $\frac{1}{2}G\rho(Ax^2 + By^2 + Cz^2) - K(x^2 - \frac{1}{2}y^2 - \frac{1}{2}z^2)$

must be constant on the free surface. This can happen only if the corresponding coefficients of the two quadratic polynomials in (566) and (567) are proportional. Therefore,

(568) $a^2(G\rho A - 2K) = b^2(G\rho B + K) = c^2(G\rho C + K).$

Now, A, B, C are determined as functions of a, b, c by (511) and its fellow travellers. Moreover, we know from (515) that

(569) $$A + B + C = 4\pi.$$

Consequently, A, B, C can be determined from (568) and (569), so a, b, c.

If the fluid rotates, there are solutions with a, b, c pairwise distinct. Let us look now for small perturbations into a spheroidal shape. Clearly, we can write

(570) $$b = c = a(1 - \epsilon), \qquad \epsilon \ll 1.$$

From (526), neglecting ϵ^2, we have

(571) $A = B = \frac{4}{3}\pi(1 - \frac{4}{5}\epsilon)$ and $C = \frac{4}{3}\pi(1 + \frac{2}{5}\epsilon).$

After division by a^2, the condition (568) can be written as

(572) $\frac{4}{3}\pi(1 - \frac{4}{5}\epsilon) - \dfrac{2K}{G\rho} = (1 - \epsilon^2)\left\{\frac{4}{3}\pi(1 + \frac{2}{5}\epsilon) + \dfrac{K}{G\rho}\right\}$

$= \frac{4}{3}\pi(1 - \frac{8}{5}\epsilon) + \dfrac{K}{G\rho} + \cdots$

if we neglect ϵ^2 and $K\epsilon$ (K also being small). The relation between K and ϵ is given to first order by

(573) $$\epsilon = \frac{45}{16\pi}\frac{K}{G\rho}.$$

To find the implications of this in the Earth-Moon system, suppose the Earth a fluid of density $5.52\,\mathrm{gm\,cm^{-3}}$. Put $M' = 7.38 \times 10^{25}$ gm, $R = 3.84 \times 10^{10}$ cm. Then $\epsilon = 8.4 \times 10^{-6}$. The radius of the Earth is about 6.4×10^8 cm, whence the difference between semiaxes is about 5.4×10^3 cm.

Note that our calculation explains why the Earth has a tidal bulge on both the side facing the Moon and the side away from the Moon. But the Earth is not a fluid mass nor is it entirely covered with water. There are frictional effects and other complications, so that the tide generally lags behind the Moon. In fact, the flattening is determined by experiment to be about $1/297 \cong 3.3 \times 10^{-3}$.

EXERCISE VIII.1. (a) Expand $(1 - 2xh + h^2)^{-1/2}$ by the binomial series into a power series in h as far as is necessary to verify the expressions given for $P_0(x)$ through $P_4(x)$. Note that $(2xh - h^2)^k$ can be expanded into a polynomial in h which contains no power of h lower than the kth.

(b) Show that the polynomial $P_n(x)$ satisfies the differential equation

(574) $$(1 - x^2)P_n'' - 2xP_n' + n(n+1)P_n = 0$$

for each nonnegative integer n and all x.

(c) Show that $\int_{-1}^{1} P_m(x)P_n(x)\, dx = 0$ for distinct integers m, n. (It can be shown further that $\int_{-1}^{1} P_n(x)^2\, dx = 2(2n+1)$.)

EXERCISE VIII.2. Prove that an oblate spheroid of uniform density can not have its boundary surface as one of its level surfaces, assuming there is no rotation.

3. Tide-raising on a ringed planet

We determine the potential at the center of a homogeneous, annular plate. Let the ring lie in the xy-plane, centered at the origin. Let the inner radius be m, the outer one n, while the density is σ per unit of surface area.

We take a point P at distance $r \ll m$ from the origin O and at colatitude ϑ. (See Figure VIII.2.) By symmetry, there is no loss of generality in supposing that P to lie in the xz-plane. Let Q be a point in the xy-plane at distance R from O and at latitude φ from the x-axis. Denote by l the distance of Q from P. Then the potential of the annulus at P is

$$
\begin{aligned}
V &= -\iint \frac{G\sigma R\,dR\,d\varphi}{l} \\
(575) \quad &= -\int_0^{2\pi}\int_m^n \frac{G\sigma R\,dR\,d\varphi}{\sqrt{(r\sin\vartheta - R\cos\varphi)^2 + R^2\sin^2\varphi + r^2\cos^2\vartheta}} \\
&= -G\sigma\int_0^{2\pi}\int_m^n \frac{dR\,d\varphi}{\sqrt{1 - 2r/R\,\sin\vartheta\cos\varphi + r^2/R^2}} \\
&= -G\sigma\sum_{k=0}^{\infty}\left\{\int_m^n \left(\frac{r}{R}\right)^k dR\right\}\left\{\int_0^{2\pi} P_k(\sin\vartheta\cos\varphi)\,d\varphi\right\}.
\end{aligned}
$$

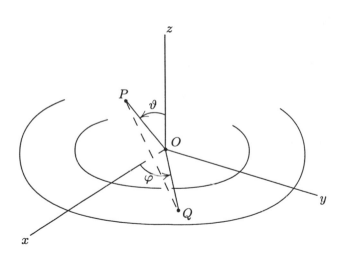

Figure VIII.2. Planetary ring(s)

Now,

$$
(576) \qquad \int_0^{2\pi} P_0(\sin\vartheta\cos\varphi)\,d\varphi = \int_0^{2\pi} d\varphi = 2\pi;
$$

$$
(577) \qquad \int_0^{2\pi} P_1(\sin\vartheta\cos\varphi)\,d\varphi = \int_0^{2\pi} \sin\vartheta\cos\varphi\,d\varphi = 0;
$$

(578)
$$
\int_0^{2\pi} P_2(\sin\vartheta\cos\varphi)\,d\varphi = \int_0^{2\pi} \tfrac{1}{2}(3\sin^2\vartheta\cos^2\varphi - 1)\,d\varphi = -\pi P_2(\cos\vartheta).
$$

Therefore, neglecting r^3,

$$(579) \qquad V = -G\sigma \left\{ 2\pi(n-m) + \pi P_2(\cos\vartheta) \left(\frac{r^2}{n} - \frac{r^2}{m} \right) \right\}.$$

But $\sigma\pi(n^2 - m^2) = M'$, the mass of the annulus, so that

$$(580) \qquad\qquad V = -\frac{GM'}{n+m} \left\{ 2 - \frac{r^2}{nm} P_2(\cos\vartheta) \right\}.$$

Now assume a small, massive, homogeneous, fluid planet to be centered at the origin. The ring causes tidal distortion, so that the planet assumes an ellipsoidal shape. As in the previous section, the tidal distortion is due to a potential

$$(581) \qquad\qquad
\begin{aligned}
V_T &= -\frac{GM'}{mn(m+n)} r^2 P_2(\cos\vartheta) \\
&= -\frac{GM'}{mn(m+n)} (z^2 - \tfrac{1}{2}x^2 - \tfrac{1}{2}y^2).
\end{aligned}$$

The role played by K in the previous section is here played by $GM'/mn(m+n)$. In consequence, the flattening is

$$(582) \qquad\qquad \epsilon = \frac{45}{16\pi} \frac{M'}{mn(m+n)\rho},$$

where ρ is the density of the planet.

EXERCISE VIII.3. Two masses M' are placed at distances R on opposite sides of a gravitating, liquid sphere of radius a and total mass M. Neglecting powers of a/R above the third, show that the liquid is deformed into a prolate spheroid, the ratio of the minor axis to the major being

$$(583) \qquad\qquad 1 - \frac{15}{2} \frac{M'}{M} \left(\frac{a}{R} \right)^3.$$

If two additional particles of the same mass M' are placed at distances R from the center on an axis at right angles to the line joining the earlier pair, show that the surface becomes an oblate spheroid, the ratio of axes being the same as before.

4. Clairaut and the variation of gravity

Before exploiting the methods of the previous sections to study the shapes of fluid ellipsoids rotating in equilibrium, we give a direct and simple derivation of an important result due to Clairaut[3] which relates the mass of the Earth to several quantities that can be determined experimentally.

Take the figure of the Earth to be a slightly oblate spheroid with semiaxes $a = b$ and $c = a(1 - \epsilon)$, where the flattening ϵ is so small that its square may be neglected. The equation of the spheroid is

$$(584) \qquad \frac{x^2 + y^2}{a^2} + \frac{z^2}{a^2(1 - 2\epsilon)} = 1,$$

or

$$(585) \qquad x^2 + y^2 + (1 + 2\epsilon)z^2 = a^2,$$

or

$$(586) \qquad r^2(1 + 2\epsilon \cos^2 \vartheta) = a^2,$$

or

$$(587) \qquad r = a(1 - \epsilon \cos^2 \vartheta),$$

where ϑ is the colatitude.

---

[3] Alexis-Claude Clairaut (1713–1765) was a child prodigy who was reading calculus by age ten. At thirteen, he presented a paper on geometry to the French Academy of Sciences. By eighteen, he had published a book and had been elected to the Academy of Sciences. In 1743, he published a work on the figure of the Earth.

In 1748, Clairaut, Lalande, and Lapaute sequestered themselves to calculate the return of Halley's comet (the comet of 1682). Lalande wrote, 'During six months we calculated from morning to night, sometimes even at meals; the consequence of which was, that I contracted an illness which changed my constitution for the rest of my life. The assistance rendered by Madame Lapaute was such that without her we should never have dared to undertake this enormous labor; in which it was necessary to calculate the distance of each of the two planets, Jupiter and Saturn, from the comet, separately for every successive degree, for 150 years.' The differential equations that they solved were not for the orbit itself but rather for the perturbations due to the two large planets. However, logarithms were probably the only calculating aids they had. The result was the prediction that the comet would reach perihelion 13 April 1749, which was in error by only 31 days. Sagan and Druyan state that this 'powerfully ... supported the Newtonian view that we live in a clockwork universe' and quote Laplace as saying 'the regularity which astronomy shows us in the movements of comets doubtless exists also in all phenomena.' (See the paper by C.W. Gear and R.D. Skeel in *A History of Numerical Computing*, Addison-Wesley, 1990.)

Assume the Earth to be a fluid mass rotating with angular velocity ω around the minor axis. Then the equilibrium of the surface requires that the potential V satisfy the equation

$$(588) \qquad V - \tfrac{1}{2}\omega^2(x^2 + y^2) = C,$$

C constant, on the free surface. Now we match with the *external* potential. If there were no rotation, the external potential could be calculated by MacCullagh's formula (473). Nevertheless, the assumed ellipsoidal shape and some calculations with formulas for the principal moments of inertia for a spheroid (444) show that the term $A + B + C - 3I$ in MacCullagh's formula is proportional to $\epsilon P_2(\cos \vartheta)$. Therefore, we set

$$(589) \qquad V = -\frac{GM}{r} - \frac{K P_2(\cos \vartheta)}{r^3},$$

where K is a constant (of order ϵ) to be determined.

To first order in ϵ,

$$(590) \qquad \frac{1}{r} = \frac{1}{a}(1 + \epsilon \cos^2 \vartheta) \quad \text{and} \quad \frac{1}{r^3} = \frac{1}{a^3}(1 + 3\epsilon \cos^2 \vartheta).$$

Therefore, the equilibrium equation is

$$(591)$$
$$\frac{GM}{a}(1 + \epsilon \cos^2 \vartheta) + \frac{K}{2a^3}(3 \cos^2 \vartheta - 1) + \frac{1}{2}\omega^2 a^2(1 - \cos^2 \vartheta) = C.$$

Because K is of order ϵ, ω^2 is also of order ϵ. Thus, it was sufficient to set $r = a$ in the second and third terms.

Because this equilibrium equation is to be an identity in ϑ, the total coefficient of $\cos^2 \vartheta$ must vanish:

$$(592) \qquad \frac{K}{a^3} = \frac{1}{3}\left(\omega^2 a^2 - \frac{2\epsilon GM}{a}\right).$$

Furthermore, the terms independent of ϑ imply that

$$(593) \qquad C = \frac{GM}{a} - \frac{K}{2a^3} + \frac{1}{2}\omega^2 a^2.$$

Therefore,

$$(594) \qquad V = -\frac{GM}{r} - \frac{a^3}{r^3}\left(\tfrac{1}{2}\omega^2 a^2 - \frac{\epsilon GM}{a}\right)(\cos^2 \vartheta - \tfrac{1}{3}).$$

The radius from the center of the Earth to a point on the surface makes an angle ϑ, the colatitude, with the z-axis. The acceleration of gravity, g, at the surface is the resultant of this central attraction and of centripetal force. If the vector of gravitational attraction makes an angle ν with the radius, then its radial component is $-g \cos \nu$. (See Figure VIII.3.) But ν is small, so that its square is negligible. Hence, at the surface,

$$g = -\frac{\partial}{\partial r}(V + \tfrac{1}{2}\omega^2 r^2 \sin^2 \vartheta)$$

$$(595) \quad = \quad \frac{GM}{r^2} + \frac{3a^3}{r^4}\left(\tfrac{1}{2}\omega^2 a^2 - \frac{\epsilon GM}{a}\right)(\cos^2 \vartheta - \tfrac{1}{3}) - \omega^2 r \sin^2 \vartheta.$$

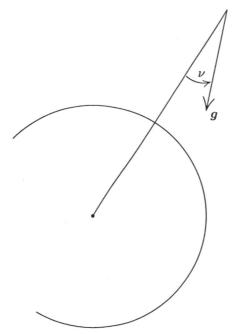

Figure VIII.3. Deflection of a plumb line

Now, $r = a(1 - \epsilon \cos^2 \vartheta)$ and ω^2 is of order ϵ. After some algebra that neglects ϵ^2, we arrive at

$$(596) \quad g = \frac{GM}{a^2}(1 + \epsilon) - \tfrac{3}{2}\omega^2 a + \left(\tfrac{5}{2}\omega^2 a - \frac{\epsilon GM}{a^2}\right)\cos^2 \vartheta.$$

Let g_e denote the value of g at the equator ($\vartheta = \pi/2$), and let ψ be the ratio of centrifugal acceleration to the acceleration of gravity at the equator (so that $\omega^2 a = g_e \psi$). Then we have

(597) $$(1 + \tfrac{3}{2}\psi)g_e = (1 + \epsilon)\frac{GM}{a^2},$$

whence $GM = (1 + \tfrac{3}{2}\psi - \epsilon)g_e a^2$ and

(598) $$g = g_e\left\{1 + (\tfrac{5}{2}\psi - \epsilon)\cos^2\vartheta\right\}.$$

(One standard formula used is

(599) $\quad g = 978.0490\,(1 + 0.0052884\cos^2\vartheta + \cdots)\,\mathrm{cm\,sec^{-2}}.$)

If g_e be measured, say by a pendulum at the equator, and if a and $\epsilon = (a - c)/a$ be observed, then the mass of the Earth can be determined from

(600) $$M = (1 + \tfrac{3}{2}\psi - \epsilon)g_e a^2/G.$$

This is *Clairaut's Theorem*.

We remarked on page 158 that the external potential takes a simple form. In fact, $I = A\sin^2\vartheta + C\cos^2\vartheta$, whence

(601) $$V = -\frac{GM}{r} + \frac{3(C - A)}{2r^3}(\cos^2\vartheta - \tfrac{1}{3}).$$

By comparison with (594), we get

(602) $$\frac{C - A}{GMa^2} = \tfrac{2}{3}(\epsilon - \tfrac{1}{2}\psi).$$

The effects of A and C can be observed in perturbations of the orbits of artificial satellites and of the Moon.

5. Poincaré's inequality for rotating fluids

Before proceeding to the study of equilibrium shapes of rotating fluid ellipsoids, we prove two general theorems about the equilibrium of rotating, homogeneous, self-gravitating fluid masses. The first of these is Poincaré's inequality, which we will prove in this section. The second, Lichtenstein's symmetry theorem, will be stated and proved in the next section.

We want to use the Divergence Theorem (sometimes known as Gauss's Theorem). Suppose that we have a region T in three-dimensional space bounded by a 'nice' surface S. Let $\partial/\partial n_e$ denote differentiation of a function across S in the direction of the unit normal n_e outward from T. Let $d\sigma$ denote the differential of surface area on S, $d\tau$ the differential of volume in T. Let W be a vector field that is continuously differentiable on T and which has a continuous extension to S. The Divergence Theorem states that

$$(603) \qquad \int_S W \cdot n_e \, d\sigma = \int_T \operatorname{div} W \, d\tau.$$

In particular, let $W = \operatorname{grad} V$, where V is a function. Then

$$(604) \qquad \operatorname{grad} V \cdot n_e = \frac{\partial V}{\partial n_e}$$

and

$$(605) \qquad \operatorname{div}(\operatorname{grad} V) = \Delta V = \frac{\partial^2 V}{\partial x^2} + \frac{\partial^2 V}{\partial y^2} + \frac{\partial^2 V}{\partial z^2},$$

the Laplacian of V. Therefore,

$$(606) \qquad \int_S \frac{\partial V}{\partial n_e} \, d\sigma = \int_T \Delta V \, d\tau.$$

Furthermore, if V is the gravitational potential produced by a mass distribution with density ρ (not necessarily constant) throughout T, then V satisfies *Poisson's equation*,

$$(607) \qquad \Delta V = 4\pi G \rho,$$

at every point of T.

To verify Poisson's equation, consider first a point P not contained in T. Let $r(P, Q)$ be the distance from P to a point Q in T. Then

$$(608) \qquad V(P) = -G \int_T \frac{\rho(Q) \, d\tau_Q}{r(P, Q)}.$$

Note that if $P = (x, y, z)$ and $Q = (a, b, c)$, then we can move the derivative inside the integral sign to get

(609)
$$\frac{\partial V}{\partial x} = \frac{\partial}{\partial x}\left\{-G\int_T \frac{\rho\, d\tau}{r}\right\}$$
$$= -G\int_T \rho\frac{\partial}{\partial x}\left\{\frac{1}{r}\right\} d\tau$$
$$= G\int_T \frac{\rho(x-a)}{r^3} d\tau$$

and

(610)
$$\frac{\partial^2 V}{\partial x^2} = G\int_T \left\{\frac{\rho}{r^3} - \frac{3\rho(x-a)^2}{r^5}\right\} d\tau.$$

Similarly,

(611)
$$\frac{\partial^2 V}{\partial y^2} = G\int_T \left\{\frac{\rho}{r^3} - \frac{3\rho(y-b)^2}{r^5}\right\} d\tau$$

and

(612)
$$\frac{\partial^2 V}{\partial z^2} = G\int_T \left\{\frac{\rho}{r^3} - \frac{3\rho(z-c)^2}{r^5}\right\} d\tau.$$

By addition,

(613)
$$\Delta V \equiv \frac{\partial^2 V}{\partial x^2} + \frac{\partial^2 V}{\partial y^2} + \frac{\partial^2 V}{\partial z^2} = 0.$$

This is Laplace's equation, satisfied at every point around which there is a little ball containing no matter.

Next, take P inside T. Suppose $P_0 = (x_0, y_0, z_0)$ a fixed point of T and ϵ a small number, and that P stays within the ball of radius ϵ centered at P_0. Suppose, moreover, that ϵ is so small that $\rho \equiv \rho(P_0)$ is a sensible approximation. Then the potential V at P can be written as

(614)
$$V = V_1 + V_2,$$

where V_1 and V_2 are the contributions from matter within and without the ϵ-sphere. We have shown that $\Delta V_2(P) = 0$. We know further that (Exercise VII.3, p. 126)

(615)
$$V_1(P) = \tfrac{2}{3}\pi\rho G(r^2 - 3\epsilon^2),$$

where r is the distance of P from P_0. It follows directly that

(616)
$$\frac{\partial^2 V_1}{\partial x^2} = \frac{\partial^2 V_1}{\partial y^2} = \frac{\partial^2 V_1}{\partial z^2} = \tfrac{4}{3}\pi\rho G,$$

whence

(617)
$$\Delta V_1(P) = 4\pi\rho G.$$

Therefore,

(618)
$$\Delta V(P) = 4\pi\rho G,$$

which is Poisson's equation.

Now suppose that T is a homogeneous, self-gravitating, noncohesive fluid mass of constant density ρ, rotating around an axis with angular velocity ω. Suppose the surface S of T is in equilibrium. Then we have *Poincaré's inequality:*

(619)
$$\omega^2 < 2\pi\rho G.$$

To prove Poincaré's inequality, note that a necessary—but not sufficient—condition for equilibrium is that the force at every point of the free surface S be directed inward. (Otherwise, pieces of the body near the surface would separate off because the fluid is not cohesive.) Suppose that U is the potential modified for rotation (4). Then

(620)
$$\frac{\partial U}{\partial n_e} > 0 \qquad \text{on } S.$$

Therefore,

(621)
$$\int_S \frac{\partial U}{\partial n_e}\, d\sigma > 0.$$

Using the Divergence Theorem (603), we obtain the inequality

(622)
$$\int_T \Delta U\, d\tau > 0.$$

But $U = V - \tfrac{1}{2}\omega^2(x^2 + y^2)$, so that

(623)
$$\Delta U = \Delta V - 2\omega^2 = 4\pi\rho G - 2\omega^2,$$

by Poisson's equation. Therefore,

$$(624) \qquad \int_T (4\pi\rho G - 2\omega^2)\, d\tau > 0.$$

The integrand is constant; therefore, $\omega^2 < 2\pi\rho G$, which is Poincaré's inequality.[4]

6. Lichtenstein's symmetry theorem

Now we prove a beautiful theorem discovered by *Lichtenstein*:[5]

> A homogeneous, self-gravitating fluid, rotating in equilibrium about a fixed axis, has a plane of symmetry perpendicular to that axis.

Suppose that T is the region of space occupied at an instant of time by the fluid and that S is the surface that bounds T. Let V be the gravitational potential:

$$(625) \qquad V(P) = -G \int_T \frac{\rho\, d\tau_Q}{r(P, Q)}.$$

Let ω be the angular velocity about the axis, which we take to be the z-axis. Choose the origin O on this axis so that the xy-plane contains the center of mass of the fluid. Embed the axes into T, and let

$$(626) \qquad U(x, y, z) = V(x, y, z) - \tfrac{1}{2}\omega^2(x^2 + y^2).$$

Then S is given by an equation

$$(627) \qquad U(x, y, z) = C,$$

for a suitable constant C.

It can be shown that V, and so U, has in T continuous first partial derivatives if it is supposed that S has a piecewise-continuous normal. We assume that S is closed.

---

[4] Crudelli proved that $\omega^2 < \pi\rho G$, sharpening Poincaré's inequality. (Y. Hagihara, *Theories of Equilibrium Figures of a Rotating Homogeneous Fluid Mass*, NASA, 1970, p. 20.) Crudelli's proof is not deep, but it would require us to spend some time developing the elementary theory of singular integrals that arise in the study of the Newtonian potential. A special case is worked out in Exercise VIII.7.

[5] L. Lichtenstein, *Mathematisches Zeitschrift*, **28**(1928), 635–640. The proof is given there for heterogeneous fluid masses.

It is clear (see the remark on p. 106) that a plane of symmetry, if it exists, must contain the center of mass of T, and that the plane must be the locus of the midpoints of all chords of T parallel to the z-axis. The strategy of the proof is to show that the midpoint locus is, in fact, a plane perpendicular to the z-axis.

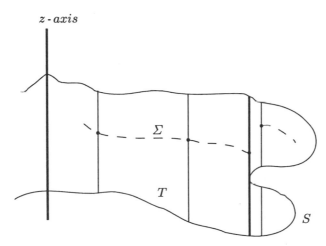

Figure VIII.4. Midpoint locus

Let Σ be the locus of all midpoints of chords of S parallel to the z-axis. (See Figure VIII.4.) If a line parallel to the z-axis has several intervals within T (including some with common endpoints at a tangency with S), take the highest midpoint. If Σ does not coincide with the plane Π through the center of mass that is perpendicular to the z-axis, then we may suppose—after making a reflection across Π if necessary—that there is a point of Σ with positive z-coordinate.[6]

The surface S is closed, so that S and T together form a compact set. Therefore, we can find a point $Q_0 = (x_0, y_0, z_0)$ inside T or on S such that z_0 is the least upper bound of the z-coordinates of all points of Σ. If $\Sigma \neq \Pi$, then there is a point on Σ with z-coordinate less than z_0.

First, suppose that Q_0 is inside T. (See Figure VIII.5.) Then Q_0 is the midpoint of a chord with endpoints (x_0, y_0, z_1) and (x_0, y_0, z_2), $z_1 > z_2$.

---

[6] Π is the plane on which $z = 0$.

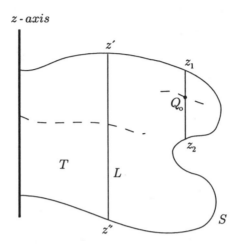

Figure VIII.5. Q_0 inside T

Now U is constant on S (or on each connected component of S, if S is not connected), so that

$$(628) \qquad V(x_0, y_0, z_1) = V(x_0, y_0, z_2).$$

Note that $\frac{1}{2}\omega^2(x^2 + y^2)$ is constant on a given vertical chord.

Let D be the projection of S onto the xy-plane. For any $\bar{P} = (\bar{x}, \bar{y}, \bar{z})$ in T or on S,

$$(629) \qquad V(\bar{x}, \bar{y}, \bar{z}) = -G\rho \int_T \frac{d\tau}{r} = -G\rho \int_D dx\, dy \int_L \frac{dz}{r(\bar{z}, z)},$$

where L is the intersection of T with the line through $(x, y, 0)$ in D parallel to the z-axis and

$$(630) \qquad r(\bar{z}, z) = \sqrt{(x - \bar{x})^2 + (y - \bar{y})^2 + (z - \bar{z})^2}.$$

If a maximal interval of L has endpoints $z' > z''$, then the definition of Q_0 implies that $z_0 \geq \frac{1}{2}(z' + z'')$. Then there is certainly an 'excess' of L that is closer to z_2 than to z_1, and we see that

$$(631) \qquad \int_L \frac{dz}{r(z_1, z)} \leq \int_L \frac{dz}{r(z_2, z)}.$$

Furthermore, there is a set of lines whose intercepts $(x, y, 0)$ have positive measure in Π for which strict inequality holds (because $\Sigma \neq \Pi$). After integration over D, we find that

$$(632) \qquad V(x_0, y_0, z_1) > V(x_0, y_0, z_2),$$

contradicting (628). Therefore, Q_0 can not lie in T, so that Q_0 must lie on S.

At a point of S, $\partial U/\partial z = \partial V/\partial z$. Either $\partial V/\partial z \neq 0$ or $\partial V/\partial z = 0$ at Q_0. We will show that neither alternative is possible if Q_0 is on S.

In the first place, suppose that $\partial V/\partial z \neq 0$ at Q_0. There is a sequence of points $(x_n, y_n, 0)$ of T converging to $(x_0, y_0, 0)$. It is clear that the midpoints of the corresponding chords L_n can not converge to z_0. (See Figure VIII.6.)

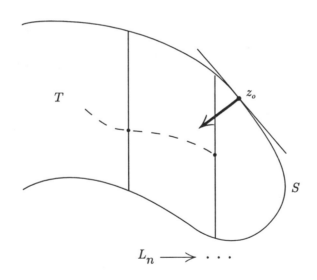

Figure VIII.6. Converging chords

In the second place, suppose that $\partial V/\partial z = 0$ at Q_0 in S. Write this as $\partial V/\partial z_0 = 0$. But

$$(633) \qquad V(x_0, y_0, z_0) = -G\rho \int_T \frac{d\tau}{r(Q_0, P)},$$

so that

$$(634) \quad \frac{\partial V}{\partial z_0}(x_0, y_0, z_0) = -G\rho \int_T \frac{\partial}{\partial z_0} \left\{ \frac{1}{r(Q_0, P)} \right\} d\tau$$

$$= -G\rho \int_D dx\, dy \int_L \frac{\partial}{\partial z_0} \left\{ \frac{1}{r(Q_0, P)} \right\} dz.$$

It is easy to see that

$$(635) \qquad \frac{\partial}{\partial z_0} \left\{ \frac{1}{r(Q_0, P)} \right\} = -\frac{\partial}{\partial z} \left\{ \frac{1}{r(Q_0, P)} \right\}.$$

Therefore, (634) becomes

$$\frac{\partial V(x_0, y_0, z_0)}{\partial z_0} = -G\rho \int_D dx\, dy \int_L -\frac{\partial}{\partial z} \left\{ \frac{1}{r(P, Q_0)} \right\} dz$$

$$(636) \qquad = -G\rho \int_D dx\, dy \left\{ \sum \left(\frac{1}{r(P', Q_0)} - \frac{1}{r(P'', Q_0)} \right) \right\},$$

where the sum is over all intervals $\{z'' \leq z \leq z'\}$ that are components of the intersection of L with T. Again, note that $z_0 \geq \frac{1}{2}(z' + z'')$. Hence, $r(P', Q_0) > r(P'', Q_0)$, and

$$(637) \qquad \frac{1}{r(P', Q_0)} - \frac{1}{r(P'', Q_0)} < 0$$

for all L. Consequently, $\partial V / \partial z_0(x_0, y_0, z_0) > 0$, contradicting the hypothesis that $\partial V / \partial z_0 = 0$.

We have shown that Q_0 can be neither in T nor on S. Therefore, Σ must coincide with Π, and this is the conclusion of Lichtenstein's Theorem.

As a corollary, we note that T must be of the form $\{(x, y, z) \mid f_1(x, y) \geq z \geq f_2(x, y)$ and (x, y) is in $D\}$. For if there were two or more 'lobes,' then Σ would lie in the higher one, which would therefore be symmetric across Π. Then a reflection in Π would move the lower lobe to the higher lobe, yielding a contradiction.

It is possible for T to be multiply connected (Saturn's rings) or to consist of disconnected pieces. The pieces must be situated along the symmetry plane, not along the symmetry axis, since there they would coalesce because of gravitation. There may be singular points on the symmetry plane. (See Figure VIII.7.)

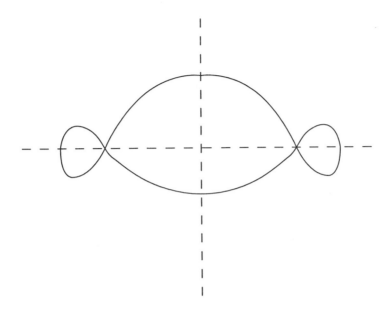

Figure VIII.7. Singular points

7. Rotundity of a rotating fluid

A rotating fluid of a given volume must have a certain 'rotundity.'

Let T denote both the region occupied by the fluid and the number expressing the volume of that region. We prove first that

(638)
$$\int_T \frac{1}{r^2} \, d\tau \le 4\pi \left(\frac{3T}{4\pi} \right)^{1/3},$$

where r is the distance from a fixed point P to a variable point in T. Note that this inequality becomes an equality if T is a ball of center P and radius $R = (3T/4\pi)^{1/3}$. Let K be the ball of center P and radius R. Call K_1 the region common to both K and T. (See Figure VIII.8.) The remainder of T will be called $T - K_1$. Then

(639)
$$4\pi \left(\frac{3T}{4\pi}\right)^{1/3} = \int_K \frac{1}{r^2} \, d\tau$$
$$= \int_{K_1} \frac{1}{r^2} \, d\tau + \int_{K-K_1} \frac{1}{r^2} \, d\tau$$
$$\geq \int_{K_1} \frac{1}{r^2} \, d\tau + \frac{1}{R^2} \int_{K-K_1} d\tau$$
$$= \int_{K_1} \frac{1}{r^2} \, d\tau + \frac{K - K_1}{R^2}.$$

Similarly,

(640)
$$\int_T \frac{1}{r^2} \, d\tau = \int_{K_1} \frac{1}{r^2} \, d\tau + \int_{T-K_1} \frac{1}{r^2} \, d\tau$$
$$\leq \int_{K_1} \frac{1}{r^2} \, d\tau + \frac{1}{R^2} \int_{T-K_1} d\tau$$
$$= \int_{K_1} \frac{1}{r^2} \, d\tau + \frac{T - K_1}{R^2}.$$

After a bit of algebra, we find that

(641)
$$\int_T \frac{1}{r^2} \, d\tau \leq 4\pi \left(\frac{3T}{4\pi}\right)^{1/3} + \frac{T - K}{R^2}.$$

However, R was chosen so that the volume K was equal to the volume T. Therefore, we arrive at the inequality (638).

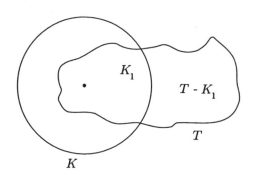

Figure VIII.8. Ball and bulge

Now, we proved in (609) that

(642)
$$\frac{\partial V}{\partial x} = -G\rho \int_T \frac{\partial}{\partial x} \left\{\frac{1}{r}\right\} \, d\tau.$$

Suppose that $Q = (a, b, c)$ is the variable point within T and $P = (x, y, z)$. Then we proved (see (610)) that

(643)
$$\frac{\partial V}{\partial x} = G\rho \int_T \frac{x - a}{r^3}\, d\tau.$$

Because $|x - a| < r$,

(644)
$$\left|\frac{\partial V}{\partial x}\right| < G\rho \int_T \frac{1}{r^2}\, d\tau \le 4\pi G\rho \left(\frac{3T}{4\pi}\right)^{1/3}.$$

Similarly,

(645)
$$\left|\frac{\partial V}{\partial y}\right|, \left|\frac{\partial V}{\partial z}\right| < 4\pi G\rho \left(\frac{3T}{4\pi}\right)^{1/3}.$$

Therefore,

(646)
$$|\operatorname{grad} V| \le 4\pi G\rho\sqrt{3} \left(\frac{3T}{4\pi}\right)^{1/3}.$$

Assume that B is a point on S where the gravitational acceleration vector ∇U reaches its maximum length. Evidently, the point B must have as much as possible of the mass between it and the axis of rotation. Considering Lichtenstein's theorem, proved in the previous section, about the symmetry of the equilibrium figure, the point B must be on the symmetry plane. The gravity vector at B is either equal to zero or directed inward; otherwise, the net outward force would cause the body to disintegrate. In particular, the centripetal force must be less than the gravitational attraction at B. Moreover, $\nabla U = \nabla V$ at B. Therefore, if B is at distance a from the axis, then

(647)
$$w^2 a \le |\nabla U| \le 4\pi G\rho\sqrt{3} \left(\frac{3T}{4\pi}\right)^{1/3}.$$

We conclude that T must lie within a cylinder coaxial with the axis of rotation and of radius a_0, where

(648)
$$a_0 = \frac{4\pi\sqrt{3}G\rho}{w^2} \left(\frac{3T}{4\pi}\right)^{1/3}.$$

This inequality is a constraint upon the spreading of the fluid into a shape that is 'too oblate.'

Note that, if T were a sphere of radius a, then (648) would imply that $\omega^2 < 4\pi\sqrt{3}G\rho$, a result weaker than Poincaré's inequality (619).

8. Ellipsoidal figures of rotating fluids

We investigate whether a homogeneous, self-gravitating fluid mass can have an ellipsoid as its equilibrium shape.

As in earlier sections, we treat the rotating fluid mass as a static one by introducing the 'fictitious' rotational potential $-\frac{1}{2}\omega^2(x^2 + y^2)$, where ω is the angular velocity around the rotation axis (taken to be the z-axis). We use the expression for the internal potential of a homogeneous, self-gravitating ellipsoid presented by (513). A surface of constant pressure is given by the equation

$$(649) \quad (\omega^2 - AG\rho)x^2 + (\omega^2 - BG\rho)y^2 - cG\rho z^2 = \text{constant.}$$

Assuming the free surface to be the ellipsoid

$$(650) \qquad \frac{x^2}{a^2} + \frac{y^2}{b^2} + \frac{z^2}{c^2} = 1,$$

the condition for this ellipsoid to be also a surface of constant pressure is that

$$(651) \qquad a^2(\omega^2 - AG\rho) = b^2(\omega^2 - BG\rho) = -c^2 CG\rho.$$

Eliminate ω^2 to get

$$(652) \qquad a^2 b^2(B - A) = (a^2 - b^2)c^2 C.$$

Thus, we can have $a = b$, whence $A = B$, and the figure is an oblate spheroid, known as a *Maclaurin spheroid*.

The value of ω is determined by the relation

$$(653) \; \omega^2 = \frac{G\rho(Aa^2 - Bb^2)}{a^2 - b^2} = 2\pi G\rho abc \int_0^\infty \frac{u\,du}{(a^2 + u)(b^2 + u)\Delta},$$

using (511) for A and its symmetrical analog for B, as well as the result of Example 44. The integral on the right side is an elliptic integral; it is positive, so that there is a real angular velocity ω.

The expression (653) holds for triaxial ellipsoids in general. If $a = b$, the case of the Maclaurin spheroid, then the integral becomes elementary.

Write $c = a\sqrt{1-e^2}$, where e is the eccentricity of the meridian ellipse. Then, setting $x = c/\sqrt{c^2 + u}$,

$$(654) \qquad \frac{\omega^2}{2\pi G\rho} = a^3\sqrt{1-e^2} \int_0^\infty \frac{u\, du}{(a^2+u)^3\sqrt{c^2+u}}$$

$$= 2(1-e^2)^2 \int_0^1 \frac{x^2(1-x^2)\, dx}{[(1-e^2)+e^2x^2]^3}.$$

Some calculation gives

$$(655) \qquad \frac{\omega^2}{2\pi G\rho} = \frac{3-2e^2}{e^3}\sqrt{1-e^2}\sin^{-1}e - \frac{3}{e^2}(1-e^2).$$

Notice that $\omega^2/2\pi G\rho$ is independent of the size of the spheroid and depends only upon its flattening.

Figure VIII.9 shows the graph of $\omega^2/2\pi G\rho$ against e, the eccentricity of the meridian ellipse. (The branch marked 'Jacobi' will be described below.)

The following properties can be derived from (655):

(1) Every oblate spheroid is a possible equilibrium form if the angular momentum of the fluid mass is suitably assigned. Note that the total angular momentum around the z-axis is $h = \frac{2}{5}Ma^2\omega$. If $r = (abc)^{1/3}$ is the radius of the ball of equal volume, then

$$(656) \qquad \frac{h^2}{GM^3r} = \frac{6}{25}\frac{\omega^2}{2\pi G\rho}(1-e^2)^{-2/3}.$$

As e increases from 0 to 1, h^2/GM^3r increases from 0 to ∞.

(2) The maximum possible value of angular velocity satisfies $\omega^2/2\pi G\rho \cong 0.2247$,[7] and this maximum occurs for $e \cong 0.9299$, when $a/c \cong 2.7198$.

(3) For each value of $\omega^2/2\pi G\rho$ less than the maximum, there are *two* possible Maclaurin spheroids, of different angular momentums. For values of $\omega^2/2\pi G\rho$ greater than the maximum, there are no Maclaurin spheroids.

(4) h^2/GM^3r is a monotonic function of e, so that there is one and only one Maclaurin spheroid for each given angular momentum.

---

[7] Compare with Crudelli's inequality. (See the footnote, p. 164.)

(5) K, the kinetic energy of rotation, satisfies

(657)
$$\frac{K}{GM^2/r} = \frac{3}{10}\frac{a^2}{r^2}\frac{\omega^2}{2\pi G\rho};$$

the dimensionless quantity on the right side begins at 0 for $e = 0$, rises to a maximum value $\cong 0.1719$ at $e \cong 0.9912$, and then decreases to 0 as the ellipsoid flattens to an infinite disk.

(6) The potential energy V satisfies

(658)
$$\frac{V}{GM^2/r} = -\frac{3}{5}\sqrt[6]{1-e^2}\frac{\sin^{-1}e}{e};$$

the dimensionless quantity on the right side increases monotonically with e from the value $\cong -0.6$ for $e = 0$ to 0 for $e = 1$.

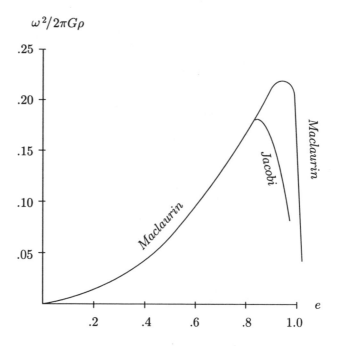

Figure VIII.9. Maclaurin and Jacobi series

As is implied by the labels in Figure VIII.9, there is another series of ellipsoids that branches off from the Maclaurin series: these are the *Jacobi ellipsoids,* and they have unequal axes. We previously have found the condition (652) for equilibrium. Using the expressions (511), etc., for A and B, we obtain

$$(659) \qquad B - A = (a^2 - b^2) \int_0^\infty \frac{abc\,du}{(a^2+u)(b^2+u)\Delta}.$$

The equilibrium condition becomes

$$(660) \qquad (a^2 - b^2) \int_0^\infty \left\{ \frac{a^2b^2}{(a^2+u)(b^2+u)} - \frac{c^2}{c^2+u} \right\} \frac{du}{\Delta} = 0.$$

If the factor $a^2 - b^2 = 0$ vanishes, then the result is a Maclaurin spheroid. But the condition for equilibrium is satisfied if the integral factor vanishes:

$$(661) \qquad \int_0^\infty \left\{ \frac{a^2b^2}{(a^2+u)(b^2+u)} - \frac{c^2}{c^2+u} \right\} \frac{du}{\Delta} = 0.$$

This can be rewritten as

$$(662) \qquad \int_0^\infty \left\{ a^2b^2 - (a^2+b^2+u)c^2 \right\} \frac{u\,du}{\Delta^3} = 0.$$

In this equation, replace (a, b, c, u) by $(\lambda a, \lambda b, \lambda c, \lambda^2 u)$; the equation is unaltered for any non-zero λ. Therefore, it represents a relation on the ratios $a : b : c$ only. Given $a : b$, then $a : c$ is determined. Moreover, rewrite the equilibrium condition as

$$(663) \qquad \int_0^\infty \left\{ \frac{1}{a^2} + \frac{1}{b^2} - \frac{1}{c^2} + \frac{u}{a^2b^2} \right\} \frac{u\,du}{\Delta^3} = 0.$$

Fix a and b and designate the integral in (663) as $f(c)$. Then

$$(664) \qquad f'(c) = \frac{2}{c^3} \int_0^\infty \frac{u\,du}{\Delta^3} > 0.$$

Because $f(c)$ is negative for very small c and positive for very large c, each pair (a, b) determines a unique c. Then, $\omega^2/2\pi G\rho$ is determined by the relation (653). It is important to notice that the root c so found is less than either a or b. For when c is very close to 0, $f(c) < 0$. However, $f(a^2b^2/(a^2+b^2)) > 0$ and $a^2b^2/(a^2+b^2)$ is less than both a and b. Therefore, Jacobi's fluid ellipsoids rotate around their shortest axis.

The angular momentum h, potential energy V, and kinetic energy T can be expressed through dimensionless ratios (with $r = \sqrt[3]{abc}$) as

$$(665) \qquad \frac{h^2}{GM^3 r} = \frac{3}{50} \frac{\omega^2}{2\pi G\rho} \frac{(a^2+b^2)^2}{abcr},$$

$$(666) \qquad \frac{V}{GM^2/r} = -\frac{3}{5} r \int_0^\infty \frac{du}{\Delta},$$

$$(667) \qquad \frac{T}{GM^2/r} = \frac{3}{20} \frac{a^2+b^2}{r^2} \frac{\omega^2}{2\pi G\rho}.$$

As Figure VIII.9 indicates, the series of Jacobi ellipsoids branches off (or *bifurcates*) from the Maclaurin spheroids. The bifurcation point can be found by writing $a = b$ and setting $c^2 = a^2(1 - e^2)$ in the integral condition (662). The integral is then elementary. Evaluations and some reductions lead to the equation

$$(668) \qquad \sin^{-1} e = (1 - e^2)^{1/2} \frac{3e + 10e^3}{3 + 8e^2 - 8e^4},$$

which has the solution $e \cong 0.8127$. For this spheroidal member of the Jacobi ellipsoids, the angular velocity is the greatest, thereafter diminishing steadily as the angular momentum increases.

As the angular momentum increases, b and c tend to equality, but with b always greater than c until the limiting form with infinite angular momentum, which consists of an infinite 'cylinder' with $a = \infty$, $b = c = 0$, and abc neither zero nor infinity. (Indeed, abc can be normalized to 1.)

The potential energy increases steadily as the ellipsoid elongates with increasing angular momentum. The kinetic energy rises to a maximum of about 0.1010. (It is 0.1006 for an ellipsoid with $a : b : c = 3.129 : 0.588 : 0.543$.)

In fact, there are *two* Jacobi series branching off the Maclaurin series, but they are analytically and physically identical and involve only an interchange of a and b. Therefore, it is necessary to consider only the branch where $a \geq b$.

EXERCISE VIII.4. Prove that there are no Jacobi ellipsoids that are slightly distorted spheres.

EXERCISE VIII.5. A homogeneous, rotating, fluid body (the Earth) is subject to the tide-raising influence of two distant bodies (the Sun and the Moon) in its equatorial plane. Find approximate figures for the fluid body when the three bodies are in a straight line (two cases) and when the lines joining the distant bodies to the central one make right angles.

EXERCISE VIII.6. A planet is in the center of a uniform, plane annulus of mass M and inner and outer radii a and b, large compared to the radius of the planet. Show that, to a first approximation, the effect of the ring on the figure of equilibrium of the planet is equivalent to that of a rotational velocity ω, where

$$(669) \qquad \omega^2 = \frac{3GM}{ab(a^2 + b^2)}.$$

EXERCISE VIII.7. A homogeneous, self-gravitating fluid rotates as an elliptic cylinder around its axis. The axes are a and b, $a > b$; the angular velocity is ω. Show that

$$(670) \qquad \frac{a - b}{a + b} = \sqrt{1 - \frac{\omega^2}{\pi\rho}},$$

verifying Crudelli's inequality[8] $\omega^2 < \pi\rho$ for this case.

---

[8] See the footnote, p. 164.

Index